Current Technologies in Vehicular Communications

George Dimitrakopoulos • George Bravos

Current Technologies in Vehicular Communications

Springer

George Dimitrakopoulos
Department of Informatics and Telematics
Harokopio University of Athens
Athens, Greece

ISBN 978-3-319-47243-0 ISBN 978-3-319-47244-7 (eBook)
DOI 10.1007/978-3-319-47244-7

Library of Congress Control Number: 2016953610

Printed on acid-free paper

This Springer imprint is published by Springer Nature
The registered company is Springer International Publishing AG
The registered company address is: Gewerbestrasse 11, 6330 Cham, Switzerland

Preface

This book covers all aspects relevant to vehicular communication technologies in one place. It classifies all relevant standards, protocols, and applications, so as to enable the reader to gain a holistic approach towards the extremely intriguing subject of vehicular communications.

The book's purpose is to become the unique place where a reader can turn to study everything that is related to vehicle to vehicle (V2V) as well as vehicle to infrastructure (V2I) technologies, classified appropriately and in a unique manner, so as to facilitate understanding.

Particular interest is placed on state-of-the-art research and development results in the field as well as research areas envisaged to attract immense research effort.

The book's main methods lie in algorithmic processes and simulation results as well as in trying to simplify all relevant technologies through a careful classification.

The book is structured as follows.

Chapter 1 provides the motivation for getting involved in the vehicular communications field, through presenting transport drawbacks and challenges.

Chapter 2 contains an extensive overview of the commonly used (and researched) standards and protocols related to V2V and V2I communications.

Chapter 3 provides a description of the context in which V2V and V2I communications operate, namely smart cities, as well as explains why smart cities are in need of novel sustainable vehicular communications. Indicative case studies give an overview of related applications in the field.

Chapter 4 focuses on Advanced Driver Assistance Systems (ADAS), presenting their main focus areas as well as including a number of case studies for exemplifying the operation of ADAS solutions.

Chapter 5 focuses on the management functionality that is researched, in the context of ADAS, focusing on the related algorithms commonly utilized.

Last, Chapter 6 gives an overview of the earlier as well as the latest trends in the field of automated and autonomous driving, providing also an outlook on the future, with some interesting perspectives for future research.

Athens, Greece George Dimitrakopoulos

Contents

List of Figures

Chapter 1
Introduction: The History of Vehicular Communications

1.1 Goals

- To make the reader familiar at a high level with the concept of transport and the requirements and trends of urban mobility.
- To distinguish between the various transport modes in terms of their requirements and potentials.
- To explain the latest trends in the various transport modes in terms of the utilization of Information and Communication Technologies (ICT).
- To introduce the reader to the book and its logic behind.

1.2 Motivation: Transportation and Its Drawbacks

Economic growth has been strongly associated with urbanization, overwhelming cities with vehicles since transportation generally and infrastructure in particular are large segments of the economy. By 2030, it is expected that around 60 % of the global population will live in urban areas[1] charting the growing contribution of cities both to the world economy and to carbon emissions. Cities are also key drivers of global energy demand and greenhouse gas emissions, accounting for around 70 % of both, according to the International Energy Agency (IEA).

This incurs a series of negative outcomes, such as:

1. Environmental/natural resource degradation (smog, polluted waterways, increased energy consumption, and CO_2 emissions).

[1]Bertaud, A. and Richardson, A.W (2004), Transit and density: Atlanta, the United States and Western Europe, Figure 17.2 on p.6, available at http://courses.washington.edu/gmforum/Readings/Bertaud_Transit_US_Europe.pdf

G. Dimitrakopoulos, *Current Technologies in Vehicular Communications*,
DOI 10.1007/978-3-319-47244-7_1

2. Socioeconomic (enormous losses of time in congestions, accidents, and degradation in life quality/deaths).
3. Technical consequences (safety compromises, accidents).[2]

These facts reveal inefficiencies related to urban transport, as identified by research communities of both public agencies and private industry.[3]

1.3 Overview of Latest Advances in Transportation Research

In response to the aforementioned challenges, this section provides a holistic view upon transportation, describing the latest advances associated with its various means, namely rail, road, air, sea, and multimodal transport.

1.3.1 Transport Mode Rail

The trends in rail vehicle development focus on lightweighting and increased use of advanced polymer composite materials and lightweight alloys. This is because rail vehicles have got heavier over the past 30 years as passengers expect a better travel experience and vehicles incorporate more ancillary equipment to enhance passenger comfort (internet access, power points, air-conditioning, noise and temperature insulation, etc.). Increased safety is also an issue of development work which, apart from the bodyshell crashworthiness integrity, also concerns new designs to mitigate terrorist action (survivability after an on board explosion) and new materials to counter fire spread. In addition, development work is undertaken on safe interiors to minimize passenger injury in case of a collision.

Advanced driver aids are also an area of development which concerns both development of automated systems to override the driver if an impending collision is likely, as well as measures to counter driver fatigue, as well as detect and enhance driver attention span (Fig. 1.1).

Other developments include the implementation of sensor technologies and electronic engine and suspension management systems (e.g., the development of mechatronic boggies, induction brakes, and energy recovery systems). Such systems aid reliability, vehicle control, vehicle efficiency, and safety and will require maintenance procedures and staff knowledge well above the current state.

---

[2] Erika Fille Legara, Christopher Monterola, Kee Khoon Lee, Gih Guang Hung, “Critical capacity, travel time delays and travel time distribution of rapid mass transit systems”, Physica A: Statistical Mechanics and its Applications, Volume 406, 15 July 2014, pp. 100–106.

[3] V. Corcoba Magaña, Muñoz-Organero, M., “Discovering Regions Where Users Drive Inefficiently on Regular Journeys”, IEEE Transactions on Intelligent Transportation Systems, vol.16, iss.1, 2015, pps. 221–234.

Fig. 1.1 Rail transport

In the railway industry the infrastructure is integral to the operation of the rail network and developments are undertaken on improved track designs (rail material, ballast, noise and vibration reduction, switching points) as well as technologies for better detection of cracks in rails, improved level crossings, and detection and warning of obstacles on the line.

Developments are also taking place in design of railway stations, improved ticketing systems, passenger flow, and passenger safety.

Furthermore, advances can be also exhibited in freight vehicle design and vehicle suspension systems, Freight vehicles traditionally have had basic suspension systems and the load-carrying capacity is much higher than a passenger vehicle. These two elements introduce significant damage on the track which in many cases is shared by passenger vehicles which increases maintenance costs for the infrastructure managers.

1.3.2 Transport Mode Road

Currents trends in road vehicle development mainly focus on three fields of action: efficiency, safety, and driving experience. Individual mobility and road transport needs to be more efficient while safety has to be increased for road users. Driving experience is a necessary factor since automotives still have to be accepted by both private and commercial vehicle users (Fig. 1.2).

Efficiency is mainly addressed by reducing the energy required by the vehicle (e.g., lightweight design, aerodynamics) and by increasing the efficiency of energy supply (e.g., vehicle electrification). The increased number of highly electrified vehicles demands new qualifications, for example, handling of high voltage systems for vehicle mechanics and mechatronics.

The field of safety can be divided into active and passive safety. The deployment of active safety measures, such as advanced driver assistance systems (ADAS), can

Fig. 1.2 Road transport

significantly reduce the number of road fatalities. However, the implementation into vehicles requires a cross-linking of domains, e.g., chassis systems with electronic systems, such as cameras and high-performance computing. Passive safety has already played a big role in the past. Nevertheless, road crashes will still occur in the future and necessitate further improvement.

Future vehicle concepts may significantly vary from today's designs. Those vehicles still have to be accepted by the users/drivers. For this reason, evaluating driving experience and driver performance plays a key role in future vehicle design and engineering projects. The required methods come from other nontechnical disciplines, such as psychology.

In addition to the already mentioned fields of actions, trade-offs require even more need for action. A higher standard of (passive) safety usually leads to higher vehicle weights and thus a reduced efficiency.

Hence, a sustainable lightweight design approach is needed. Active safety, especially the introduction of highly automated or even autonomous vehicles, and driver experience lead to the necessity of evaluating driver performance in various situations, especially driver reactions during automatic vehicle interventions. It should not be forgot that vehicles still need to be attractive (private cars) and utilizable (commercial vehicles) while efficiency is improved, leading to innovative approaches regarding energy and propulsion management.

Finally, the introduction and implementation of ITS and intermodal mobility concepts may generate new vehicle classes, e.g., purposely designed car sharing vehicles or vehicle concepts for highly agglomerated areas (e.g., vehicle class L7e).

1.3.3 *Transport Mode Air*

Continuing growth in air travel generates increasing volumes of surface traffic traveling to and from airports. Over the past 50 years the average annual growth rate of the aviation industry was 5 %. Despite the recent economic recession, this trend is expected to continue over the next decades. However, existing infrastructure developments cannot keep in pace with the current demand. At almost all major airports, terminals are problematic bottlenecks for traffic. One approach to avoid problems and delays for people and freight, alternatives to existing air travel are proposed by facilitating the ability to access airports by high occupancy airport access modes like rail. Taking into account the increase in air traffic and the expansion of urban areas around airports, intermodality between air and rail transport modes has become increasingly important. Up until now, intermodality is focused in shipment of freight across different transport modes. However, many researches argue that true and measurable efficiency gains from intermodality will come regarding passenger intermodality, which has not gained attention in all parts of the world (Fig. 1.3).

In order to mitigate air traffic congestion, management of all large airports are very keen to transfer the intermodal "revolution" beyond the air cargo business, in order to integrate airports with other transport modes.

The term "intermodal transportation" refers to a system that connects separate transportation modes—roads, aviation, maritime, railway—that allows a passenger to complete a journey using more than one mode. Intermodal integration provides passengers not only with the ability to connect to an extended transportation network, but also with a safe and efficient (i.e., "seamless") transfer between the various modes. An intermodal connection in air transportation, for instance, might involve a passenger arriving at the airport by local rail service, flying from there to

Fig. 1.3 Air transport

another airport, and finally transferring to a private hotel shuttle service. The transfer from one mode to the next is conducted at the so-called "intermodal terminals." Since air passengers need to get to and from the airport, they generally rely on more than one mode of transportation during their journey. Consequently, trips that include at least one section covered by air are usually "intermodal" journeys.

The close interaction between different means of transportation requires the transportation chain to be treated as a continuous end-to-end process. This implies that problems arising either on the air- or the landside of an airport cannot be handled independently. On the other hand, the transportation system as a whole may benefit from improvements (e.g., better information coordination or increased connectivity) in one part of the chain.

The European market for air transportation is the second most important aviation market in the world. Regarding intermodality between air and rail transport perhaps the most successful case is the Frankfurt airport. More specifically, the airport suffers from severe airside capacity problems, and—during peakhours—from landside congestion problems on nearby highways and access routes. An integrated railway station was established in 1972. In 1999 this regional train station has been expanded by another station specifically designed for long-distance trains and providing access to the German Intercity Express (ICE) high-speed network. In 2002, a railway corridor from the airport to the densely populated Rhine-Ruhr area was opened. As of 2008, over 160 high-speed rail services are offered each day. The airport serves as a mainline station in the trans-European high-speed rail network and is thus easily accessible. A joint venture "AIRail" was launched by the airport operator (Fraport), the main airline at the airport (Lufthansa), and the railway company (Deutsche Bahn) to support the intermodal integration at the airport.

Services comprise a check-in area at the rail station that has a direct link to the airport's automated baggage handling system. Additionally, AIRail services are provided on high-speed trains to the cities of Cologne and Stuttgart. Passengers check in their baggage at the respective train station and receive flight boarding passes. Trains are assigned flight numbers, with passengers flying business or economy class receiving the corresponding seats on the trains (1st and 2nd class). Passengers only pay a single bill.

1.3.4 Transport Mode Waterway/Sea

The current state of the art in shipping and marine transportation is characterized by a mixture of used technologies and procedures. More than in other transport modes in international shipping there is a mixture of conventional technologies and even traditional procedures. Modern containerships with increasing dimensions and equipped with latest navigation, timing, and positioning technologies and devices including enhanced decision support systems using sophisticated sensor and enhanced data management, uninterruptedly monitored by company-owned fleet operation centers (FOC) especially for safety purposes and especially in coastal waters also monitored and supported by vessel traffic services (VTS) are sailing across the globe. Such most modern and big ships share waterways with other smaller ships, equipped rather outdated and operated by applying traditional techniques and methods (Fig. 1.4).

Fig. 1.4 Waterway/sea transport

"Unmanned vessels" is a topic that is speedily reascending the priority agendas of various maritime administrations, and indeed the IMO. The IMO should be well prepared for a time when, rather than if, unmanned vessels become a reality. In order to do so, however, significant areas need to be addressed in a sufficient manner.

In order to improve safety and efficiency of shipping, the latest development in international shipping are mainly driven by IMO's e-Navigation initiative that calls for the implementation of latest information and communication technologies (ICT) to support the human operators on board with additional information and advice. There is an obvious trend for introducing and implementing more new electronic devices and to improve data collection and processing for enhanced opportunities for traffic management in terms of coordination and even discussing comprehensive traffic control from a supervising monitoring and control center.

Consequently, one must first define what the potential areas of concerns could be. Rather than reinventing the wheel, the IMO should consider the case of other high risk industries—particularly nuclear, aero- and astronautics, and automotive industries—that have, to different extents, experienced the challenges brought about by the advent of advanced control systems, unmanned technology, and robotics.

1.3.5 Intermodal Transport

The Transport White paper[4] issued by the European Commission aims to "by 2020, establish the framework for a European multimodal transport information, management and payment system." Multimodal transport refers to the transportation of goods

[4] European Commission. (2011). White Paper on transport. Roadmap to a single European transport area—towards a competitive and resource-efficient transport system. Illustrated brochure. European Commission, Directorate General Mobility and Transport. http://ec.europa.eu/transport/themes/strategies/2011_white_paper_en.htm.

Fig. 1.5 Intermodal transport

and people by two or more modes of transport (such as road, rail, air, inland waterway and sea). Through the introduction of ICT the use of multimodal transport can become easier since it can provide the user with an up to date integrated service (Fig. 1.5).

Potter[5] discussed transport integration and an integrated transport policy which appears to be similar to the White Paper (goal no. 8) and argued that a common definition is not possible but that it should include the following:

- Locational Integration: being able to easily change between transport modes—services connecting in space.
- Timetabling Integration: Services at an interchange connect in time.
- Ticketing Integration: Not needing to purchase a new ticket for each leg of a journey.
- Information Integration: Not needing to enquire at different places for each stage of a trip (or that different independent sources are connected to appear seamless to users).
- Service Design Integration: That the legal, administrative and governance structures permit/encouraging integration.
- Travel Generation Integration: Integrating the planning of transport with the generators of travel (particularly integration with land use planning).

Concerning operation and organization of intermodal transport one of the main barriers is the incompatible infrastructure that results from the fact that most of the

[5] Potter, S. (2010). Transport integration—an impossible dream? Universities Transport Studies Group Annual Conference, 5–7 January 2010, University of Plymouth

existing transport infrastructure has been designed to serve the national rather than the international economy. Such limitations have resulted in something called "cross-border bottlenecks." Another problem is the lack of comprehensive standards regarding infrastructure design, power supplies, traffic management and data.

Another barrier to intermodal transport includes not only the absence of a common EU standard including privacy standards but also the lack of technical standards. The lack of technical standards results in not being implemented on a regional scale let alone on a national and international scale.

Multimodal transport information systems should make it easier to use more than one mode of transport. However, some systems on the market are too difficult to use and instead of making the journey easier it could make it more complex. For instance some ticket machines at train stations do not provide enough information or sometimes too much information.

Moreover, the market is growing very quickly and it could therefore be difficult for the customer to be familiar with all the different applications on the market. This is also something which is still affecting the introduction of transportation management applications.

The conclusion is that despite various efforts to improve the ICT systems it is still far from achieving the goal of establishing a framework for a European multimodal transport information, management and payment system. Incompatible infrastructure is one important barrier which prevents integration. Another problem is that many different actors are involved who have different requirements. On an international level, the lack of legislation is the most important barrier and that those who provide data have to pay a duty, which of course could be regarded as a disincentive.

1.4 Road Transport: Utilization of ICT in Vehicles: Intelligent Transport Systems (ITS)

This section takes as a basis the advances described in the previous section and elaborates on road transport. As mentioned also above, there has been an emergence of innovative, cost-effective cooperative mobility and automated driving solutions improving energy efficiency, individual safety and the effectiveness of public and freight transport. These initiatives together form the cornerstone of Intelligent Transport Systems (ITS).[6,7,8,9]

---

[6] Ahmad, A.; Arshad, R.; Mahmud, S.A.; Khan, G.M.; Al-Raweshidy, H.S., "Earliest-Deadline-Based Scheduling to Reduce Urban Traffic Congestion", IEEE Transactions on Intelligent Transportation Systems, Volume:15, Issue: 4, 2014, pps. 1510–1526

[7] Alam, K.M, Saini, M., El Saddik, A., "Toward Social Internet of Vehicles: Concept, Architecture, and Applications", IEEE Access, vol.3, 2015, pps. 343–357

[8] G. Dimitrakopoulos, P. Demestichas, "Intelligent Transportation Systems based on Cognitive Networking Principles", IEEE Vehicular Technology Magazine (VTM), March 2010

[9] Festag, A., "Cooperative intelligent transport systems standards in europe", IEEE Communications Magazine, Volume: 52, Issue: 12, 2014, pps: 166–172

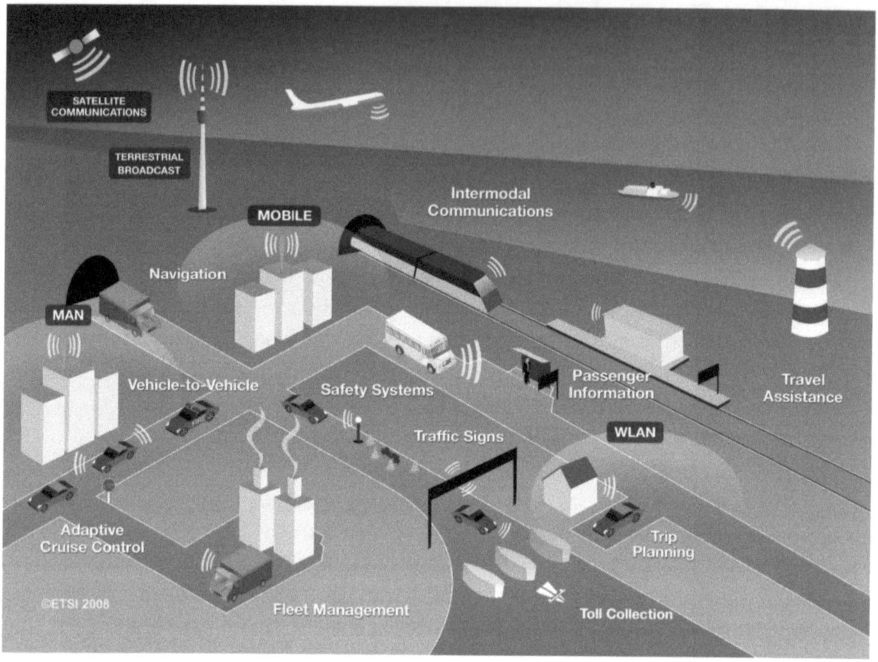

Fig. 1.6 General ITS vision (ETSI, 2008)

By enabling vehicles to communicate with each other via Vehicle to Vehicle (V2V) communication as well as with roadside base stations via Vehicle-to-Infrastructure (V2I) communication, ITS can contribute to safer and more efficient roads.[10,11,12]

The general ITS vision is shown on the following figure (Fig. 1.6):

From a complementary perspective, advanced safety systems for both, vehicles and pedestrians are established as a most important service requirement in the transportation field, for numerous countries.

Despite the—still partial—establishment of traffic management/safety systems, there is still a long way to go for maximizing transportation efficiency and safety. Some of the causes that may be identified are the following:

- The traffic conditions that should be handled by the transportation infrastructure may frequently change. Traffic changes can be sudden or recurring. Sometimes they may be due to the occurrence of emergencies. Traffic is to be assessed in real-time, and communicated to the drivers, in such a way it can be taken into account.

[10] Hayes, J., "Calling all cars", Engineering & Technology journal, vol. 9, iss. 3, 2014, pps. 58–61

[11] Kolosz, B.W. et al, "A Macroscopic Forecasting Framework for Estimating Socioeconomic and Environmental Performance of Intelligent Transport Highways", IEEE Transactions on Intelligent Transportation Systems, vol.15, iss. 2, 2014, pps. 723–736

[12] El Mouna Zhioua, G., Tabbane, N., Labiod, H., Tabbane, S., "A Fuzzy Multi-Metric QoS-Balancing Gateway Selection Algorithm in a Clustered VANET to LTE Advanced Hybrid Cellular Network", IEEE Transactions on Vehicular Technology, vol. 64, iss.2, 2015, pps. 804–817

Traffic patterns resulting from a learning process should add more accuracy to the messages communicated to the drivers.

- Sudden changes into the traffic are often the direct cause of accidents, again the transport infrastructure is expected to communicate these changes in a timely fashion in order to decrease the risk of accident and therefore casualties which would result in an even worse traffic condition.
- Legacy traffic assessment and management systems are mainly centralized. This means that, in principle, they are unsuitable either for adapting, in short time scales, to context changes, or for supporting cooperation of the relevant services. Currently, the collection of context information, the solution of optimization problems and the application of reconfiguration decisions is an off-line process, applied in medium (or long) time scales.
- Intelligence embedded in vehicles is still at a low level in terms of their communication capability with external entities (i.e., other vehicles and/or objects of the transportation infrastructure) and there is no assessment in the vehicle of the overall security status that would rely on a correlation of the global traffic condition and the vehicle and driver behaviors.
- There is no direct correlation between the status and the crossing lights and the traffic occurring at the corresponding segments. Safety level there is minimal and just the result of assumption or rules like "if the light turns red, car will stop and therefore 3 s after pedestrian will be able to cross."
- There is no direct communication between the traffic infrastructure and the cars about which directions are to avoid or to follow. The drivers are not aware of the traffic conditions in real time and have no mean therefore to behave or adapt accordingly.

1.5 Conclusions

Mobility as a concept is indispensably linked to economic growth. However, this has incurred severe drawbacks that apparently affect all transport modes. To eliminate these drawbacks, the world has been moving towards the adoption of ICT-enabled solution for connected vehicles and objects of the transportation infrastructure, exchanging information and tackling sudden or recurring situations.

The above give birth to numerous concepts, all revolving around vehicular communications. It is thus of great research interest to study the "why" and the "how" of transport connectivity, as well as to investigate whether it brings any benefits to making transport sustainable.

1.6 Review Questions

Question 1.1:
How can we prove that mobility is an indispensable part of human initiative in the twenty-first century?

Question 1.2:
What is multimodal transport, compared to individual/conventional transport modes?

Question 1.3:
What is V2V/V2I/V2X communications?

Question 1.4:
How can transport systems become "intelligent", exploiting the advent of ICT?

Question 1.5:
What are the basic research areas in ITS that provide the motivation for further research and development efforts?

Chapter 2
Vehicular Communications Standards

2.1 Goals

- To present all available and emerging standards related to V2V and V2I communications.
- To focus on emerging standards for V2V/V2I.
- To present concrete use cases so as to familiarize the reader with the V2V/V2I logic.
- To set the scene for gaining knowledge on the various standards' drawbacks, so as to work on new topics to eliminate them.

2.2 Introduction

As already mentioned, the main motivation for vehicular communication systems is safety and eliminating the excessive cost of traffic collisions. According to World Health Organization (WHO), road accidents annually cause approximately 1.2 million deaths worldwide,[13] one-fourth of all deaths caused by injury. Also about 50 million persons are injured in traffic accidents. If preventive measures are not taken road death is likely to become the third leading cause of death in 2020 from ninth place in 1990. A study from the American Automobile Association (AAA) concluded that car crashes cost the United States $300 billion per year.[14]

[13] M. Peden; Richard Scurfield; D. Sleet; D. Mohan; et al. "World report on road traffic injury prevention" (PDF). World Health Organization. Retrieved April 15, 2016

[14] "Crashes Vs. Congestion—What's the Cost to Society?" (PDF). American Automobile Association. Retrieved April 15, 2016.

G. Dimitrakopoulos, *Current Technologies in Vehicular Communications*,
DOI 10.1007/978-3-319-47244-7_2

In general, V2I communications have been implemented based on numerous standards, such as IEEE 802.11n, DSRC, and Infrared techniques. They have been widely deployed for road charging applications but the infrastructure cost makes the cost/benefit calculation challenging, demanding significant investment overhead. Further, Wide Area Networking (WAN) technologies such as 2G/GPRS/EDGE, 3G/UMTS/HSPA/HSPA+, and 4G/LTE have also been used for vehicle to back office communication, but these suffer from location accuracy which could be improved by secondary mechanism such as GPS.

On the other hand, the concept of (mostly neighboring) vehicles communicating with each other has been the subject of research and development initiatives for many years. However, the level of adoption of V2V techniques in modern vehicles has only recently started to increase and it is still far below satisfactory levels.

Lately, through the connectivity available for vehicles, vehicles have started getting connected to the internet, giving birth to several applications that fall in the realm of V2B (Vehicle-to-Business) communications.

Last but not least, the increasingly rising utilization of smart devices has produced a new generation of mobile apps so that a driver can be connected to his/her vehicle remotely.

In this respect, this chapter aims at outlining the standards that are being used in V2X communications, emphasizing on the advantages and the drawbacks of each one of them.

2.3 Wireless Access for Vehicular Environments (WAVE) and Its Migration Towards IEEE 802.11p

Wireless Access for Vehicular Environments (WAVE) is an approved amendment to the IEEE 802.11 standard. WAVE is also known as IEEE 802.11p. WAVE is required to support the Intelligent Transportation Systems (ITS) applications in the short-range communications. The communication between vehicles (V2V) or between the vehicles and the roadside infrastructure (V2I) is relied on the band of 5.9 GHz (5.85–5.925 GHz).[15] With the equipment installed in the car and on the road, WAVE supplies the real-time traffic information, improves the safety of the transportation, and reduces the traffic congestion. It also benefits for the transport sustainability.

In 1992, United States started to research the Dedicated Short Range Communication (DSRC). It is the wireless communication protocol for the vehicles. United States, Europe, and Japan are the main countries of research and application for DSRC. From 2004, the concentration of DSRC has been migrating to the IEEE 802.11 standard group. At first DSRC is based on the IEEE 802.11a, which focus on the low overhead operations. DSRC standard is based on the Wireless

[15] Stephan Eichler, "Performance Evaluation of the IEEE 802.11p WAVE Communication Standard", in Proceedings of Vehicular Technology Conference, 2007, pp.2199–2203

Fidelity (Wi-Fi) architecture.[16] However, in order to support high-speed moving vehicle and simplify the mechanisms for communication group, IEEE working group dedicate more efforts on the WAVE, which is the core of the DSRC. WAVE ensures the traffic information collection and transmission immediate and stable, and keeps the information security.

Besides the IEEE 802.11p, WAVE also contains the standard of IEEE 1609, which is the upper layer standard. IEEE 1609 completes the WAVE by its sub-detail standards, for instance, IEEE 1609.2 standard is responsible for the communication security; IEEE 1609.3 standard covers the WAVE connection setup and management.[17] IEEE 1609.4 standard that is based on the IEEE 802.11p Physical (PHY) layer and Medium Access Control (MAC) layer supplies operation of high-level layers across multiple channels.

In general, standards-based vehicular networking for V2V communication has been so far implemented to a great extent, based on IEEE 802.11p,[18,19] which inherits several of the IEEE 802.x family characteristics, including simplicity and distributed medium access control mechanisms. It is at an early stage of adoption however and though it does meet the requirement for minimal infrastructure investment, it suffers from reliability, resilience to interference and stability problems, as well as faces the "fax machine problem"—it's only any good if you can communicate with a second party that has similar equipment.

Despite its limited applicability, several applications based on the IEEE 802.11p standard are on the market, but some projects are testing yet with few vehicles. A large number of companies, car manufacturers, and universities are involved in those projects, and we may see appearance in our cars the first applications in the next few years. The application can be separated into three aspects.

2.3.1 Safety-Oriented

Most of applications are safety related, but these applications need real-time constraints that the IEEE 802.11p is not able to provide itself. So some extensions of the amendment are needed to allow the use of safety applications.[20]

---

[16] D. Jiang, L. Delgrossi, "IEEE 802.11p: Towards an International Standard for Wireless Access in Vehicular Environments", in Proceedings of Vehicular Technology Conference, 2008, pp.2036–2040

[17] Task Group p, "IEEE 1609.3-2007 WAVE Networking Services", IEEE Computer Society, 2007

[18] F Bai, H Krishnan, Reliability analysis of DSRC wireless communication for vehicle safety applications, in Proceedings of the IEEE Intelligent Transportation Systems Conference (ITSC 2006), (Toronto). 17–20, September 2006

[19] A Vinel, 3GPP LTE versus IEEE 802.11p/WAVE: which technology is able to support cooperative vehicular safety applications? IEEE Wireless Commun. Lett. 1(2), 125–128 (2012)

[20] Bohm, A.Jonsson, "Position-Based Data Traffic Prioritization in Safety-Critical, Real-Time Vehicle-to-Infrastructure Communication", CERES (Centre for Res. on Embedded Syst.), Halmstad Univ, Halmstad, Sweden,

The communication could be based on the point to point or multipoint. It also demands the low latency requested by the real-time communication. Car-to-car communication (C2C) can be used to provide a global view of the traffic that the driver could not be able to have by himself. For example, by exchanging of information such as position and speed, a driver can see on a screen in his car all vehicles around. This is very useful if the weather prevents a good visibility, like fog or rain, and in a turn or at an intersection. A driver can also be advertised of a traffic jam or a traffic collision. This is also very useful especially if the driver have a bad visibility. For an emergency vehicle, because it has to arrive at the destination without delay, it can broadcast a message to the cars around it and make a place for itself.

Car-to-infrastructure communication (C2I) can be used, for example, to allow an emergency vehicle to preempt a red light on its way, and then have green light all along its path, or at the intersection, the traffic light sends the light information to the cars that are in its communication scope. It assists drivers better know about the conditions of the intersection to avoid traffic collision.

2.3.2 *Traffic Control-Oriented*

Some other applications are not related to the safety, but by exchanging information about position we can have a global view about the density of the traffic and used it to regulate the traffic. For example, the traffic jam advertiser, enumerated for safety purpose, is also a traffic control-oriented application in a way that the user knows about a traffic jam further and then can choose an other way.

We can also imagine a "smart red light" that could collect information about number of cars waiting and how long time they have been waiting, and then change its status based on that.

The infrastructure can also supply the localization map for the drivers and make a suggestion of appropriate path to the destination and avoid traffic jam. The Electronic Toll Collection (ETC) has been applied in some Europe countries. ETC charges the road price for reducing the congestion. The system can recognize the car by car's identification by the equipment based on the WAVE technology without stopping the cars. The antenna installed on the car can communicate with the on-board equipment, which is on the car.

2.3.3 *User Comfort-Oriented*

Some previous applications could be also in this section, such as the traffic jam advertiser or the smart red light, because they can avoid the driver to wait too long time in a traffic jam or at a red light. But the comfort-oriented applications are more service that the users could enjoy themselves in their cars like download movies or music or upload some documents to their friends. Actually having access to the Internet can summarize comfort applications.

Some research initiatives are in-going for that, but for some obvious reasons the IEEE 802.11p is not design for that. First of all having always a path to an access point for the Internet is almost impossible, because of the high mobility of vehicles, which should be routers. There is also a big problem of security in a way that it is not possible to trust any routers on the path. So having the Internet now in our vehicles by using the IEEE 802.11p amendment is not a really good solution and using other technology like the 3G is still better.

2.4 IEEE 1609

The IEEE 1609 family of standards defines the following parts:

- Architecture
- Communication model
- Management structure
- Security mechanisms

Physical access for high-speed (<27 Mb/s), short-range (<1000 m), and low latency wireless communications in the vehicular environment.

The primary architectural components defined by these standards are the On Board Unit (OBU), Roadside Unit (RSU), and WAVE interface.

The IEEE 1609.3 standard covers the WAVE connection setup and management. The IEEE 1609.4 standard sits right on top of the IEEE 802.11p and enables operation of upper layers across multiple channels, without requiring knowledge of PHY parameters. The standards also define how applications that utilize WAVE will function in WAVE environment. They provide extensions to the physical channel access defined in WAVE.

This is shown in Fig. 2.7.

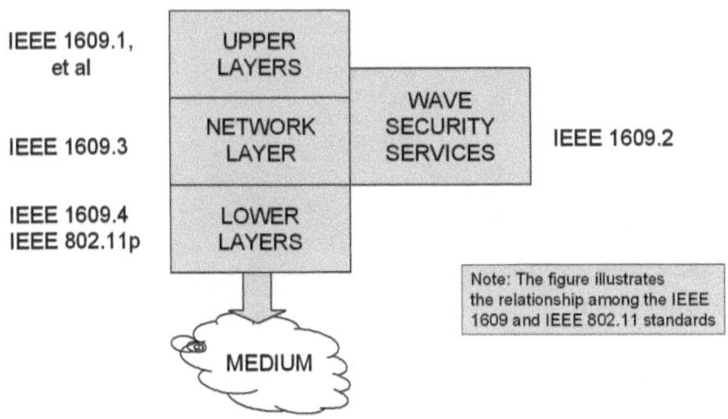

Fig. 2.7 IEEE 1609

2.5 SAE J2735

Another standard that is commonly used in vehicular communications and, in particular, V2V communications is the J2735, Dedicated Short Range Communications (DSRC) Message Set Dictionary, maintained by the Society of Automotive Engineers (http://www.sae.org). This SAE Standard specifies a message set, its data frames, and data elements specifically for use by applications intended to utilize the (DSRC/WAVE) communications systems.

Although the scope of this Standard is focused on the message set and data frames of DSRC, it specifies the definitive message structure and provides sufficient background information for the proper interpretation of the message definitions from the point of view of an application developer implementing the messages according to the DSRC standards.

It supports interoperability among DSRC applications through the use of standardized message sets, data frames, and data elements. The message sets specified in J2735 define the message content delivered by the communication system at the application layer and thus defines the message payload at the physical layer. The J2735 message sets depend on the lower layers of the DSRC protocol stack to deliver the messages from applications at one end of the communication system (OBU of the vehicle) to the other end (a roadside unit). The lower layers are addressed by IEEE 802.11p, and the upper layer protocols are covered in the IEEE 1609.x series of standards.

The message set dictionary contains:

15 Messages
72 Data Frames
146 Data Elements
11 External Data Entries

The most important message type is the basic safety message (often informally called “heartbeat” message because it is constantly being exchanged with nearby vehicles). Frequent transmission of “heartbeat” messages extends the vehicle’s information about the nearby vehicles complementing autonomous vehicle sensors. Its major attributes are the following:

- Temporary ID
- Time
- Latitude
- Longitude
- Elevation
- Positional Accuracy
- Speed and Transmission
- Heading
- Acceleration
- Steering Wheel Angle
- Brake System Status
- Vehicle Size

The other kinds of messages are the following:

A la carte message—composed entirely of message elements determined by the sender, allowing for flexible data exchange.

Emergency vehicle alert message—used for broadcasting warnings to surrounding vehicles that an emergency vehicle is operating in the vicinity.

Generic transfer message—provides a basic means to exchange data across the vehicle-to-roadside interface.

Probe vehicle data message—contains status information about the vehicle to enable applications that examine traveling conditions on road segments.

Common safety request message—used when a vehicle participating in the exchange of the basic safety message can make specific requests to other vehicles for additional information required by safety applications.

2.6 LED-Enabled Visible Light Communications (IEEE TG 802.15.7)

Light emitting diodes (LEDs) constitute a well-established choice for light sources in display and illumination applications. LEDs combine the advantages of high brightness and low power as well as low heat dissipation and longer life span compared to conventional incandescent lamps. Moreover, LED lamps are an important candidate for road illumination, traffic signs, and vehicle head lights.

However, according to medium to long-term research EU roadmaps, technology will enable the enhancement of any real-world object (such as traffic signs, road lights, and vehicle head lights), even the simplest, with ICT capabilities. These smart objects will be equipped with sensors, actuators, and embedded processors and will need to adopt an open networked architecture. In this respect, considering that LEDs can also be modulated at relatively high speeds, this offers the intriguing possibility of realizing the illumination or display functionality and at the same time of transmitting data. This concept is usually referred to as VLC,[21,22] and provides to the overlying applications increased reliability, significantly reduced energy footprint, interference-free transmission, cost efficiency (LED lights already installed for various applications), as well as easy integration and interoperability.

However, since almost all vehicles dispose LED lights, it would be very easy and cost-efficient to utilize those LEDs for additional purposes, such as for offering fast,

[21] O. Bouchet et al, "Visible-light communication system enabling 73 Mb/s data streaming 2010 IEEE Globecom Workshops", GC'10, art. no. 5700092, pp. 1042–1046.

[22] T.Komine et al, "Basic Study on Visible-Light Communication using Light Emitting Diode Illumination," Proc. of the 11th Int. Symposium on Personal, Indoor and Mobile Radio Communications (PIMRC 2000), London, US, pp. 1325–1329, 2000.

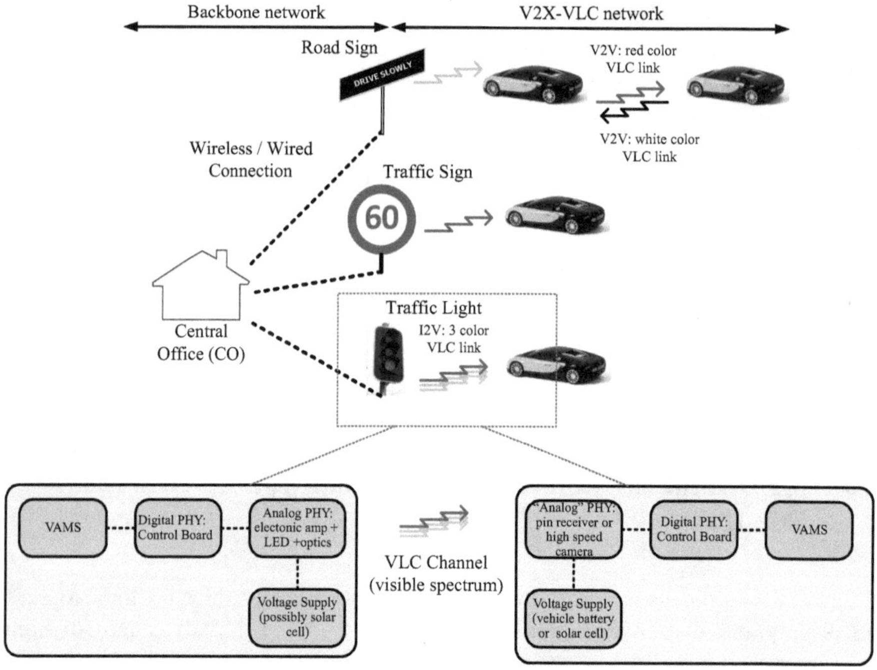

Fig. 2.8 VLC as a standard for V2X communications

reliable, and energy-efficient information to the driver, complementing and cooperating with other available solutions, as will be explained in the sequel. As such, LED-VLC seems a natural candidate for realizing V2X communications.[23]

Figure 2.8 illustrates the basic concepts associated with a system relying on VLC. The system is decomposed into a backbone network connecting the central office (CO) to the various access points of the network (traffic lights, road signs, etc.). The CO is responsible for coordinating and managing the information exchange between the vehicles and the infrastructure. The backbone network can be implemented using existing wire-line or wireless technologies (fiber-to-the-x, ADSL, RF links, etc.). The second part of the network is based on VLC technology and consists of the various V2I and V2V connections.

The VLC links can be used for:

- Downstream connection from the traffic lights to the vehicles using the LEDs of the lights as a means of transmitting data. All three LED colors must be used here (red, green, yellow).
- Downstream connection from road and traffic signs to the vehicles. The connections are realized through the sign's single color LED.

[23] Binti Che Wook et al, "Visible light communication with LED-based traffic lights using 2-dimensional image sensor", CCNC 2006, 1, art. no. 1593024, pp. 243–247 (2006).

- Upstream connection between the vehicle and the various access points. This can be realized using the vehicle's white LED lights.
- Upstream and downstream connection between the vehicles. These are supported with the LED front and break lights. During daytime it may be interesting to consider IR LEDs embedded in the vehicle lights.
- Upstream connection from the vehicle to the access point through either visible or IR LEDs.
- Supporting decision-making in the "Vehicle Autonomic Management System" (VAMS).

Overall, VLC is a valid candidate for complementing current solutions in the world of transportation, through offering (1) high reliability, (2) low infrastructure cost, (3) very low carbon emissions, and (4) resilience to interference.

2.7 Bluetooth

Bluetooth technology is a wireless communications technology that is simple, secure, and can be found almost everywhere. You can find it in billions of devices ranging from mobile phones and computers to medical devices and home entertainment products. It is intended to replace the cables connecting devices, while maintaining high levels of security. Automotive applications of Bluetooth technology began with implementing the Hands-Free Profile for mobile phones in cars. The development is coordinated by the Car Working Group (CWG) and is ongoing ever since 2000 by implementing different profiles and new features. The key features of Bluetooth technology are ubiquitousness, low power, and low cost. The Bluetooth Specification defines a uniform structure for a wide range of devices to connect and communicate with each other.

When two Bluetooth-enabled devices connect to each other, is the so-called pairing. The structure and the global acceptance of Bluetooth technology means any Bluetooth-enabled device, almost everywhere in the world, can connect to other Bluetooth-enabled devices located in proximity to one another.

Connections between Bluetooth-enabled electronic devices allow these devices to communicate wirelessly through short range, creating ad hoc networks commonly known as piconets. Piconets are established dynamically and automatically as Bluetooth-enabled devices enter and leave radio proximity, meaning that you can easily connect whenever and wherever it's convenient for you. Each device in a piconet can also simultaneously communicate with up to seven other devices within that single piconet and each device can also belong to several piconets simultaneously. This means the ways in which you can connect your Bluetooth devices is almost limitless. There are applications that even do not require a connection establishment. It may be enough if the Bluetooth device's wireless option is set to "visible" and "shown to all," because fixed positioned Bluetooth access points may detect the movement of the Bluetooth device from one AP to another AP. This technology can easily be used for measuring the traffic flow.

A fundamental strength of Bluetooth wireless technology is the ability to simultaneously handle data and voice transmissions, which provides users with a variety of innovative solutions such as hands-free sets for voice calls, printing and fax capabilities, and synchronization for PCs and mobile phones, just to name a few.

The range of Bluetooth technology is application specific. The Core Specification mandates a minimum range of 10 m or 30 ft, but there is no set limit and manufacturers can tune their implementations to provide the range needed to support the use cases for their solutions.

Range may vary depending on class of radio used in an implementation:

- Class 3 radios—have a range of up to 1 m or 3 ft
- Class 2 radios—most commonly found in mobile devices—have a range of 10 m or 33 ft
- Class 1 radios—used primarily in industrial use cases—have a range of 100 m or 300 ft

Bluetooth technology operates in the open and unlicensed industrial, scientific, and medical (ISM) band at 2.4–2.485 GHz, using a spread spectrum, frequency hopping, full-duplex signal at a nominal rate of 1600 hops/s. The 2.4 GHz ISM band is available and unlicensed in most countries. The most commonly used radio is Class 2 and uses 2.5 mW of power. Bluetooth technology is designed to have very low power consumption. This is reinforced in the specification by allowing radios to be powered down when inactive.

Bluetooth technology's adaptive frequency hopping (AFH) capability was designed to reduce interference between wireless technologies (such as WLAN) sharing the 2.4 GHz spectrum. AFH works within the spectrum to take advantage of the available frequency. This is done by the technology detecting other devices in the spectrum and avoiding the frequencies they are using. This adaptive hopping among 79 frequencies at 1 MHz intervals gives a high degree of interference immunity and also allows for more efficient transmission within the spectrum. For users of Bluetooth technology this hopping provides greater performance even when other technologies are being used along with Bluetooth technology.

The newest Bluetooth Technology is Bluetooth 4.0 called Bluetooth Smart (Low Energy) Technology. While the power efficiency of Bluetooth Smart makes it perfect for devices needing to run off a tiny battery for long periods, the most important attribute of Bluetooth Smart is its ability to work with an application on the smartphone or tablet you already own. Bluetooth Smart wireless technology features:

- Ultra-low peak, average and idle mode power consumption
- Ability to run for years on standard coin-cell batteries
- Low cost
- Multi-vendor interoperability
- Enhanced range

In automotive industry the primary usage of Bluetooth connects hands-free car systems which help drivers focus on the road. Another special usage is health monitoring, e.g., people with diabetes can monitor their blood glucose levels by using a

Bluetooth glucose-monitoring device paired with the car. Also in-vehicle intelligent interfaces may provide, e.g., vehicle-related technical information to the driver via a Bluetooth channel.

In V2I systems Bluetooth can be used to provide communication channel between the car and the traffic signal systems. Nowadays several manufacturers offer Bluetooth capable traffic control devices. It is capable for privileging the public transport at the intersections or measuring the traffic and pedestrian flows with the help of the electronic devices installed with Bluetooth radio (such as smartphones, tablets, and navigation units). These systems detect anonymous Bluetooth signals transmitted by visible Bluetooth devices located inside vehicles and carried by pedestrians. This data is then used to calculate traffic journey times and movements. It reads the unique MAC address of Bluetooth devices that are passing the system. By matching the MAC addresses of Bluetooth devices at two different locations, not only the accurate journey time is measured, privacy concerns typically associated with probe systems are minimized.

2.8 2G and 3G Mobile Communication Infrastructures

The most wide-spread mobile (cellular) network technology is GSM (Global System for Mobile communication). GSM was designed principally for voice telephony, but a range of bearer services was defined (a subset of those available for fixed line Integrated Services Digital Networks, ISDN), allowing circuit-switched data connections at up to 9600 bits/s. The technology behind the Global System for Mobile communication (GSMTM) uses Gaussian Minimum Shift Keying (GMSK) modulation, a variant of Phase Shift Keying (PSK) with Time Division Multiple Access (TDMA) signaling over Frequency Division Duplex (FDD) carriers. Although originally designed for operation in the 900 MHz band, it was soon adapted also for 1800 MHz. The introduction of GSM into North America meant further adaptation to the 800 and 1900 MHz bands. Over the years, the versatility of GSM has resulted in the specifications being adapted to many more frequency bands to meet niche markets.

At the time of the original system design, this rate compared favorably to those available over fixed connections. However, with the passage of time, fixed connection data rates increased dramatically. The GSM channel structure and modulation technique did not permit faster rates, and thus the High Speed Circuit-Switched Data (HSCSD) service was introduced in the GSM Phase 2+.

During the next few years, the General Packet Radio Service (GPRS) was developed to allow aggregation of several carriers for higher speed, packet-switched applications such as always-on internet access. The first commercial GPRS offerings were introduced in the early 2000s. Meanwhile, investigations had been continuing with a view to increasing the intrinsic bit rate of the GSM technology via novel modulation techniques. This resulted in Enhanced Data-rates for Global Evolution (EDGE), which offers an almost threefold data rate increase in the same bandwidth. The combination of GPRS and EDGE brings system capabilities into the range covered by the International Telecommunication Unions IMT-2000 (third generation) concept, and some manufacturers and network operators consider the EDGE networks to offer third generation services.

In 1998, the ETSI (European Telecommunications Standards Institute) General Assembly took the decision on the radio access technology for the third generation cellular technology: wideband code-division multiple access, W-CDMA, would be employed. A dramatic innovation was attempted: a partnership project was formed with other interested regional standards bodies, allowing a common system to be developed for Europe, Asia, and North America. The Third Generation Partnership Project (3GPP) was born.

The Third Generation mobile cellular technology developed by 3GPP—known variously as Universal Mobile Telecommunications System (UMTS), Freedom of Mobile Multimedia Access (FOMA), 3GSM, etc., is based on wideband code division multiple access (W-CDMA) radio technology offering greater spectral efficiency and higher bandwidth than GSM. UMTS was originally specified for operation in several bands in the 2 GHz range. Subsequently, UMTS has been extended to operate in a number of other bands, including those originally reserved for Second Generation (2G) services. The UMTS radio technology is direct-sequence CDMA, each 10 ms radio frame is divided into 15 slots.

As a development of the original radio scheme, a high-speed download packet access (HSDPA, offering download speeds potentially in excess of 10 Mbit/s) and an uplink equivalent (HSUPA, also sometimes referred to as EDCH) were developed. Collectively the pair are tagged HSPA, and permit the reception of multimedia broadcast/multicast, interactive gaming and business applications, and large file download challenging traditional terrestrial or satellite digital broadcast services and fixed-line broadband internet access. The radio frames are divided into 2 ms subframes of 3 slots, and gross channel transmission rates are around 14 Mbit/s.

3GPP's radio access undergoes continuous development and the "long-term evolution" (LTE) exercise aims to extend the radio technology

2.9 4G/5G-D2D

2.9.1 Concept Overview

Given the diverse performance requirements of a wide spectrum of vehicular networking applications and the huge cost of deployment of specialized road infrastructure, research has been currently moving towards the investigation of the benefits of exploiting the existing/emerging mobile communication standards (LTE—X2 interface, and most importantly 5G Device-to-Device—D2D) as suitable mechanisms for delivering automotive applications, with a focus on autonomous driving (AD). It is envisaged that next generation mobile technologies (4G/4G+, 5G), including D2D networking and very low latency communications, will constitute alternative technology solutions to 802.11p.

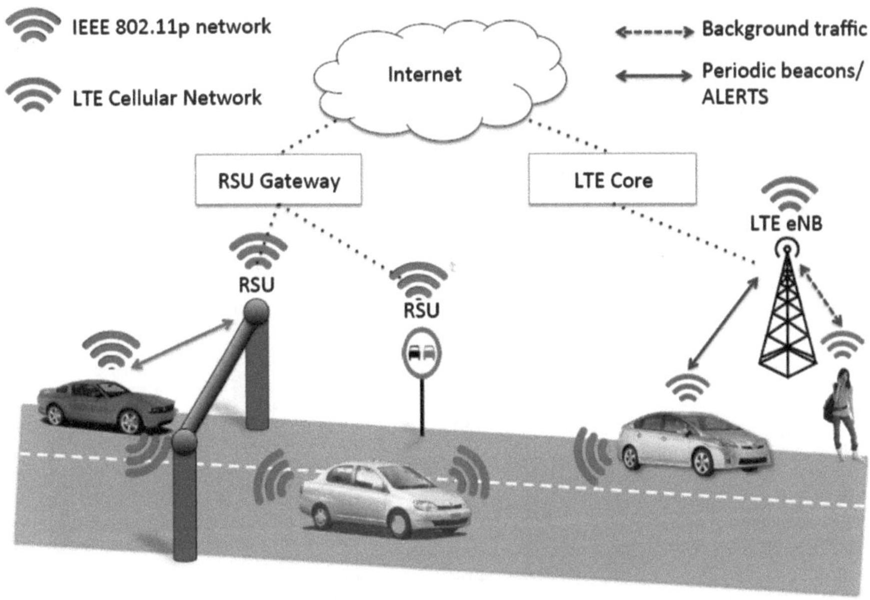

Fig. 2.9 Exploitation of 4G/5G mobile communication infrastructures in V2V and V2I

The main idea of this trend is to exploit emerging wireless standards to leverage Mobile Network Operators (MNOs) existing telecommunication infrastructures and (network) data, to enhance intelligence on the move for providing novel Advanced Driver Assistance Systems solutions.

In this respect, the framework depicted on Fig. 2.9 can exploit 4G/5G MNO infrastructure/date for V2V/V2I communications as an alternative to conventional approaches based on the utilization of costly Roadside Infrastructure/Units (RSU) and IEEE 802.11p. This solution might be able to promise multidimensional advantages since it promises (a) reduced latency, (b) increased reliability, (c) a more efficient and pervasive market penetration model, and (d) cost-efficiency.

A 2nd illustration of this framework, more detailed one, is presented in the following figure (Fig. 2.10). As shown, the framework should utilize various data "sources," aggregates the collected information through a Data Acquisition, Pre-processing and Fusion (DAPF) module, processes it on the basis of Cognitive Decision-Making (CDM) functionality, and provides as output directives to drivers to support them in accident avoidance and to mitigate the consequences of collisions.

Yet, ALL message transmissions foreseen to take place can be realized FULLY through the existing MNO telecom infrastructures instead of needing to build costly roadside infrastructures. The latter can of course be additionally exploited when and where available, for further enhancing road safety, but it is not a necessary condition for the success of such solutions.

The following subsections present, in detail, the information sources that such solutions use, the information itself, as well as some example operational scenarios that showcase its effectiveness.

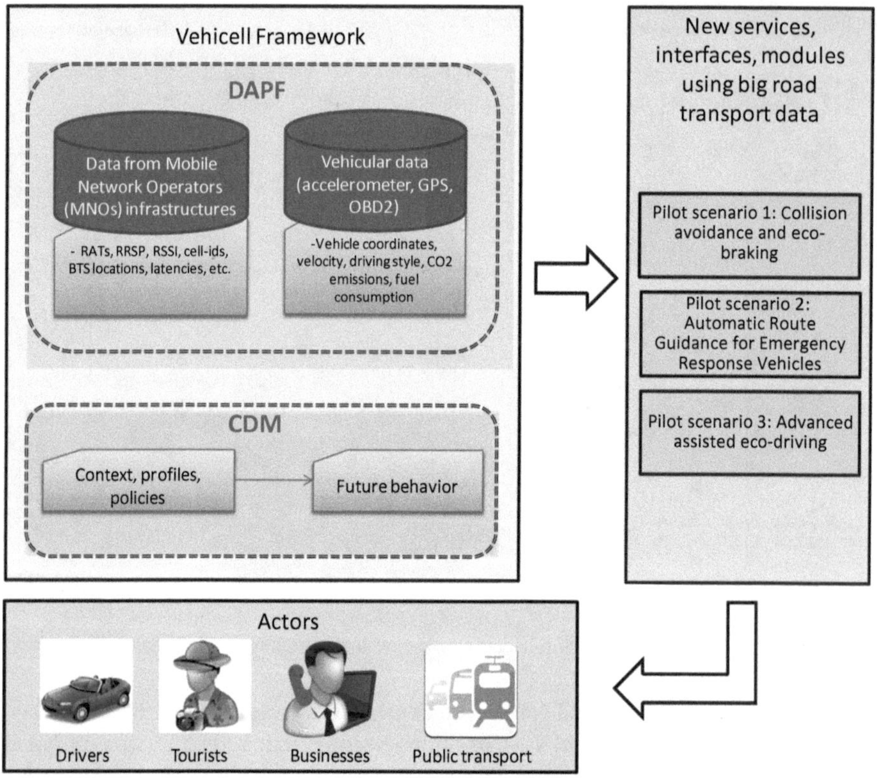

Fig. 2.10 Exploitation of mobile communication infrastructures in detail

2.9.2 *Information Sources*

As already mentioned, the fundamental novelty of this approach lies in the utilization of the MNOs telecom infrastructures, for any "message" transmission, instead of any other costly V2I technologies (requiring roadside infrastructures) and/or unreliable V2V technologies. To do so, the following data "sources" are utilized:

- A mobile smartphone (inside the vehicle) and/or an on-board device (if available)
- The vehicle itself (via an OBD-II device)
- MNO-related data

These 3 data sources can provide significantly useful information for drivers, with minimum costs, reduced latency, and high reliability, as will be shown in the sequel.

2.9.3 *Example Data to Be Aggregated*

Example data that will be collected from the abovementioned data sources are the following:

Smartphone/Tablet and/or On-Board Device (ADAS)

- MNO-related info, such as (a) Cell-id, LAC, and Radio Access Technology (RAT) currently utilized, (b) Received Signal Strength Indicator (RSSI) and/or Reference Signal Received Power (RSRP) from the serving Base Stations and/or from the neighboring ones, etc.
- Information from motion sensors, environmental sensors, and position sensors, such as accelerometers, gravity sensors, gyroscopes and rotational vector sensors, barometers, photometers, and thermometers, orientation, and magnetometers sensors.
- Location information (from GPS) such as latitude and longitude.

Vehicle/OBD-II

- Current and average speed, acceleration, throttle/boost, coolant temperature
- Timings (0–60 Km/h, 0–100 Km/h, 0–1000 m, etc.)
- Current and average CO_2 emissions (trip, overall)
- Current and average consumption (trip, overall)
- Tank level, etc.

MNO Data

- Location of Base Stations (i.e., GPS coordinates)
- RAT supported per Base Station (GSM, UMTS, HSPA, HSPA+, 4G)

2.9.4 *Processing and Outcomes*

The processing of the aforementioned information is made on the basis of smartphone applications that constantly provide the RSSI/RSRP level of the phone from the x-nearest BTSs, where $x>3$ (often $x\approx 10$). This information is extremely useful if combined with additional data provided by the MNO, such as the location of BTS/Node-Bs, the RAT (GSM, UMTS, LTE, etc.), as with the help of triangular and multi-angular calculations and GPS data (if and whenever available-considering users reluctance to utilize an always-GPS ON application due to high battery consumption), it can result in the specification of coordinates of the cell phone with very high accuracy, its velocity, etc.

Moreover, this information can pave the way for significant improvements in the provision of fast, tailor-made information which the driver is capable of processing in changing conditions, in the sense that the efficiency of this and therefore its level of "automation" depends on the RAT that is locally and currently provided by the MNO. This "progressive" procedure is further justified in the scenarios presented below.

Cell phone accelerators can act complementarily to the above information, if we consider that the acceleration/deceleration of a vehicle can exploit cognitive principles and machine learning techniques, in order to result in extracting the driver's profile and proactively identifying a forthcoming emergency, judging from the driver's reactions, which will be provided through the cell phone accelerator. The driver's profile can be used in adding further enhancements to the directives provided.

Finally, specific data extracted from the vehicle through OBD2 and sent to the cell phone or on board device can also be exploited in providing innovative nature assistance to drivers.

2.9.5 Benefits of Framework

This approach can bring about significant advantages, with respect to safe and connected automation in road transport, compared to existing solutions, for all stakeholder involved (drivers, citizens, public authorities, and businesses). These advantages can be summarized as follows:

- *>50 % reduced latency*. This approach operates with significantly lower latencies (from 10 to 20 ms) compared to existing vehicle connectivity solutions (e.g., IEEE 802.11p—>60 ms) and thus can guarantee for faster decision-making and, in return, increased active safety for drivers[24]
- *20–30 % increased reliability and robustness*. The framework will bring a revolution to current road transport automation solutions since it will be able to provide a support for accurate, stable, reliable, and proactive tailor-made directives (assistance) to drivers. Reliability is higher than that of existing approaches (e.g., IEEE 802.11p,[25,26])
- *30–40 % increased cost-efficiency*. The framework is inherently cost-efficient since its main idea from its conception was to use existing infrastructures (namely telecommunication infrastructure), whilst exploiting their benefits (reduced latency, increased reliability), in order to provide innovative ADAS to drivers without the costs for road infrastructure
- *10–20 % resulting reduced energy footprint*. The CDM along with its decision-making support process can constitute a seminal move towards enhancing Green Driving Support Systems, through providing "greener" directives that will result in lower CO_2 emissions by at least 10 % compared to current conditions, which will positively impact the European society as a whole

[24] Hameed Mir and Filali, "LTE and IEEE 802.11p for vehicular networking: a performance evaluation", EURASIP Journal on Wireless Communications and Networking 2014, 2014:89

[25] KA Hafeez, L Zhao, Z Liao, BN Ma, Performance analysis of broadcast messages in VANETs safety applications, in Proceedings of the IEEE Global Telecommunications Conf. GLOBECOM 2010, Miami, FL, 6–10 December 2010

[26] F Bai, H Krishnan, Reliability analysis of DSRC wireless communication for vehicle safety applications, in Proceedings of the IEEE Intelligent Transportation Systems Conference (ITSC 2006), (Toronto). 17–20, September 2006

- *30 % increased security, privacy and confidentiality.* The solution is protected against jamming and tapping through the utilization of the mobile communication infrastructure mechanisms, compared to current vehicular networking approaches (e.g., IEEE 802.11p,[27,28,29])
- *Easy integration (availability).* With such a solution, drivers, businesses, and public service providers will have the possibility to communicate with each other and share useful information. One of the major challenges of public service providers and local authorities is the need to manage the multiple interfaces to the different legacy and newly introduced systems. This solution will overcome this by introducing a unified frontend to the system having a built in capability to support and integrate legacy deployed solutions and thus introduce the real added value for the operators and decision makers in adoption of the solution from the business perspective. This is particularly easy due to the utilization of existing user equipment (smartphones)

2.9.6 Operational Scenarios

This subsection describes some indicative scenarios that showcase the efficiency of the solutions that exploit mobile communication infrastructures in providing novel ADAS solutions.

2.9.6.1 Scenario 1: Collision Avoidance

Scenario 1 envisages a faster, more reliable, and more secure collision avoidance use case compared to today's solutions, as illustrated in Fig. 2.11.

A cell phone inside a vehicle (blue vehicle) is currently located within an LTE service area, and a potential emergency incident (e.g., sudden brake) takes place close to it. The vehicle involved in the incident (white vehicle) can notify the nearest eNB and the eNB can inform the blue vehicle accordingly, with the communication taking place through the X2-AP protocol, over the X2 interface designed for LTE. The overall required latency is <20 ms, this being appropriate for most of today's available as well as future vehicular applications.

The resulting advantages of such low latency are obvious since even emergency braking can be activated. The emergency messages will have a structure compliant with the ETSI standards Decentralized Environmental Notification Message

[27] G Araniti et al, "LTE for vehicular networking: a survey", IEEE Commun. Mag. 51(5), 148–157 (2013)

[28] HY Kim, DM Kang, JH Lee, TM Chung, A performance evaluation of cellular network suitability for VANET. World Academy of Science, Engineering and Technology, International Science Index 64, 6(4), 1023–1026 (2012)

[29] A Vincl, 3GPP LTE versus IEEE 802.11p/WAVE: which technology is able to support cooperative vehicular safety applications? IEEE Wireless Commun. Lett. 1(2), 125–128 (2012)

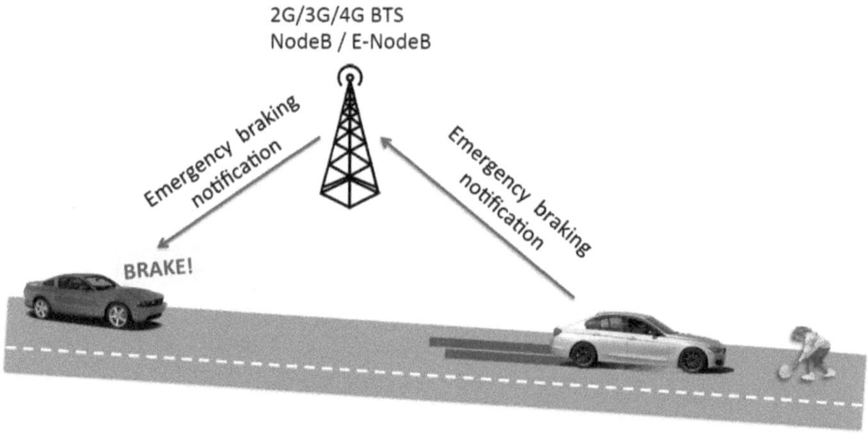

Fig. 2.11 Scenario 1—collision avoidance and eco-braking

(DENM)[30] and Cooperative Awareness Message (CAM).[31] Therefore the applications layers for the emergency braking functions will be identical, whether the messages are transmitted only through mobile communication systems (i.e., LTE). Moreover, the absence of roadside infrastructure costs renders this solution attractive.

The aforementioned scenario is even more challenging in the case of a 5G service area, where the potential D2D communication (without even the communication through any base station) will support even lower latencies (<10 ms), paving the way for progressively autonomous driving applications. On the other hand, in the case that the cell phones in vehicle A and vehicle B are located in a UMTS (3G) service area, latency drops down to >60 ms. This can be acceptable for some fundamental vehicular applications. Moreover, considering the existence of a GSM service area, the required latency would be >600 ms on average. This latency is calculated stressing the necessity of data (re)transmission within the same cell, which is needed for minimizing the required Round Trip Time (RTT). However, such high latencies are inappropriate for most of the vehicular applications since they prohibit any substantial real-time emergency management.

In general, such solutions also explore all the potential combinations among the current RATs (e.g., GSM–UMTS, and LTE-UMTS) since neighboring vehicles may be served through versatile RATs.

[30] ETSI TS 102 637-3—Decentralized Environmental Notification Message, (http://www.etsi.org/deliver/etsi_ts/102600_102699/10263703/01.01.01_60/ts_10263703v010101p.pdf), accessed February 2015.

[31] ETSI TS 102 637-2—Cooperative Awareness Message (http://www.etsi.org/deliver/etsi_ts/102600_102699/10263702/01.02.01_60/ts_10263702v010201p.pdf), accessed February 2015.

2.9.6.2 Scenario 2: Automated Route Guidance for Emergency Response Vehicles

From data regarding traffic speed of individual vehicles and traffic concentration, it will be possible to generate a real-time traffic map, with a high degree of accuracy. Compared to current mobile apps (e.g., www.waze.com), solutions of this kind usually promise higher reliability, since it will guarantee service provision without necessitating a community participation). Using intelligent algorithms, it is possible to determine the fastest route to a destination, utilizing actual times, rather than estimates that ignore delays caused by irregular traffic (a frequent occurrence) or unforeseeable events (accidents). By plotting the processed data onto a map, it is possible to display:

- Real-time Average speed of traffic on a road, including (a) on an individual segment of road, (b) at an intersection, and (c) on individual road lanes.
- Predict changes in those traffic speeds, based on traffic in the broader area
- The fastest route to a destination based on current traffic speeds and precise estimations of future traffic speeds. The time to destination will be calculated based on the driver's "driving profile."

Furthermore, the software is able to identify which vehicle in a fleet can reach a destination in the shortest time, taking into consideration actual situational awareness, rather than broad, inaccurate estimations. Such route guidance can one day lead to fully automated response services and can be utilized by unmanned taxiing companies for efficient passenger transport, with a profound impact on eco-driving, by providing significant improvements in fuel economy.

2.9.6.3 Scenario 3: Advanced Assisted Eco-Driving

The 3rd scenario envisages the utilization of the accelerometer and of the in-vehicle *OBD-II*, so as to devise eco-driving directives. In particular, the accelerometer and the *OBD-II* can provide several data on the current fuel consumption, the CO_2 emissions, the road vehicle condition, and the driver profile (driving style). This information is usually aggregated by a software module that is able to aggregate large amounts of heterogeneous data and provides a twofold outcome:

- Extract the most appropriate (re)route for the vehicle dynamically, depending on the traffic (real-time estimation through RSSI/RSRP/GPS) and the expected CO_2 emissions and fuel consumption.
- Gather long-term statistical data leading to knowledge and experience in order to extract an overall eco-driving profile of the driver and provide information to insurance companies.

2.10 ETSI and CEN Standards for V2X Communications

Two EU organizations (ETSI and CEN) have been performing research towards identifying and specifying new standards for vehicular communications. In this respect, they have recently announced[32] connected car standards that pave the way for V2V (vehicle to vehicle) and V2I (vehicle to infrastructure) throughout Europe. More and more connectivity is being added to vehicles and as a result their attack surface is increasing, as is the list of potential implications of cyber attack against vehicles.

The standards are: the specification of Cooperative Awareness Basic Service—EN 302 637-2, and the specification of Decentralized Environmental Notification Basic Service—EN 302 637-3. They define the message sets needed for running Cooperative ITS safety critical applications. Published as Technical Specifications in Release 1 of ETSI ITS, the ENs have been prepared taking into account feedback from Plugtests interoperability testing workshops organized by ETSI for the industry, as well as feedback from implementation. They were developed under Mandate 453 of the European Commission.

The Cooperative Awareness Service enables the exchange of information between road users and roadside infrastructure, providing each other's position, dynamics, and attributes. Road users may be cars, trucks, motorcycles, bicycles, or even pedestrians, while roadside infrastructure equipment includes road signs, traffic lights, or barriers and gates. Awareness of each other is the basis for several road safety and traffic efficiency applications. This is achieved by regular exchange of information from vehicle to vehicle (V2V), and between vehicles and roadside infrastructure (V2I and V2I) based on wireless networks. EN 302 637-2 specifies the syntax and semantics of the Cooperative Awareness Message (CAM) and provides detailed specifications on the message handling.

EN 302 637-3 defines the Decentralized Environmental Notification (DEN) Basic Service that supports road hazard warning. The Decentralized Environmental Notification Message (DENM) contains information related to a road hazard or an abnormal traffic condition, including its type and position. Typically for an ITS application, a message is disseminated to ITS stations that are located within a geographic area through direct vehicle-to-vehicle or vehicle-to-infrastructure communications in order to alert road users of a detected and potentially dangerous event. At the receiving side, the message is processed and the application may present the information to the driver if it is assessed to be relevant. The driver is then able to take appropriate action to react to the situation accordingly.

[32] europa.eu/rapid/press-release_IP-14-141_en.htm

2.11 Conclusions

This chapter has gone through the available standards for vehicular communications. Let it also be noted it was intentionally avoided to distinguish between V2V standards and V2I standards because many of them, especially the newest ones, have been designed so as to serve both types of communication.

Overall, from an implementation point of view, the problem with V2V communications so far is that they suffer from the fax-machine problem, i.e., all vehicles on route should have the technology implemented. Imagine, e.g., an incident where 3 cars are involved, where 2 of them are enabled with any of the technologies that support V2V communications, whereas the 3rd one is not. The accident would not be avoided.

On the other hand, the problem with V2I communications is that usually they require high installation costs. As a result, only a few cities can afford to have everywhere sensors and/or internet-enabled objects of the transportation infrastructure.

In conclusion, researchers are now trying to put into effect hybrid standards, in that they should enable both V2V and V2I communications.

2.12 Review Questions

Question 2.1:
What are the most commonly utilized standards for V2V communications?

Question 2.2:
What are the most commonly utilized standards for V2I communications?

Question 2.3:
What are the barriers for the adoption of V2V standards and what for V2I standards?

Question 2.4:
Why and how mobile communication infrastructures can be a candidate for V2X communications?

Question 2.5:
How does 5G-D2D promise low latencies in vehicular communications?

Question 2.6:
Is LED-VLC a better standard than IEEE 802.11p?

Chapter 3
Sustainable Mobility in Smart Cities: Traffic Assessment, Forecasting, and Management

3.1 Goals

- To familiarize the reader with the concept of smart cities and smart city operations.
- To present the technological state of the art and challenges related to the concept of urban sustainable mobility.
- To make the reader understand the various problems that cities face in terms of transportation inefficiencies.
- To present concrete use cases on traffic assessment, forecasting, and management.

3.2 Urban Transportation Inefficiencies

There is an increasing demand for mobility across the globe. As a result, most large cities are overcrowded with vehicles and face unpleasant daily outcomes, including traffic congestion, pollution, and accidents. Current mobility strategies are inefficient, leading to enormous losses of time, safety compromises, pollution, and degradation in the quality of life, as identified by research community of both public agencies and private industry.[33,34] Moreover, given the current energy source mix, the above inefficiencies lead to a huge waste of nonrenewable fossil energy, indicating the necessity for more efficient and safer mobility.

---

[33] G. Dimitrakopoulos, P. Demestichas, "Intelligent Transportation Systems based on Cognitive Networking Principles", IEEE Vehicular Technology Magazine (VTM), March 2010.

[34] Toppeta, D., The Smart City Vision: How Innovation and ICT Can Build Smart,"Livable", Sustainable Cities, 2010, The Innovation Knowledge Foundation. Available from http://www.thinkinnovation. org/file/research/23/en/Toppeta_Rcport_005_2010. Pdf

G. Dimitrakopoulos, *Current Technologies in Vehicular Communications*,
DOI 10.1007/978-3-319-47244-7_3

In response to these challenges, there has been an emergence of innovative, cost-effective cooperative mobility and automated driving solutions improving energy efficiency, individual safety, and the effectiveness of public and freight transport. These initiatives together form the cornerstone of Intelligent Transport Systems (ITS).[35,36,37,38] By enabling vehicles to communicate with each other via Vehicle-to-Vehicle (V2V) communication as well as with roadside base stations via Vehicle-to-Infrastructure (V2I) communication, ITS can contribute to safer and more efficient roads.[39,40,41]

However, despite numerous advances over the past few years, there are still plenty of efficiency gains to be had, such as:

(a) Better information to the traveler: Real-time, accurate, and tailored information provision to the driver, especially when it originates from multiple sources and is associated with large amounts of data, is essential.
(b) Faster Response: Solutions that will assist the driver in effectively handling sudden or unforeseen situations (Advanced Driver Assistance Systems—ADAS).
(c) Reducing the cost of deployment: Centralized cost, as necessitated by infrastructure-based systems, is difficult to justify when the benefit is distributed.
(d) Better Reliability: Reliability of communication infrastructure and resilience to interference.
(e) Better Cooperation and Integration: In-vehicle intelligence, connectivity with other vehicles, and coordination among heterogeneous technologies.

3.3 Smart Cities and Smart City Operations (SCOs)

3.3.1 Basic Definitions

There is no unique definition of the term "smart city." Instead, there have been several attempts to provide descriptive definitions of the term. As such, according to Giffinger et al.,[42] smart city is a city well performing in a forward-looking way in economy,

[35] http://www.ibm.com/smarterplanet/us/en/smarter_cities/overview/, accessed February 26th, 2015

[36] http://www.hitachi.com/products/smartcity/download/pdf/whitepaper.pdf, accessed March 16th, 2015

[37] BIS, The Smart City Market: Opportunities for the UK, 2013, Department of Business, Innovation and Skills.

[38] Naphade, M., Banavar, G., Harrison, C., Paraszczak, J., Morris, R., "Smarter Cities and Their Innovation Challenges", IEEE Computer, Volume: 44, Issue: 6, 2011, pps. 32–39

[39] http://www.ibm.com/smarterplanet/us/en/smarter_cities/overview/, accessed February 26th, 2015

[40] http://www.hitachi.com/products/smartcity/download/pdf/whitepaper.pdf, accessed March 16th, 2015

[41] BIS, The Smart City Market: Opportunities for the UK, 2013, Department of Business, Innovation and Skills

[42] Giffinger, R., Fertner, C., Kramar, H., Kalasek, R., Pichler-Milanović, N., & Meijers, E., "Smart Cities: Ranking of European Medium-Sized Cities", Vienna, Austria: Cen-tre of Regional Science

people, governance, mobility, environment, and living, built on the smart combination of endowments and activities of self-decisive, independent, and aware citizens. Likewise, the authors of "Foundations for Smarter Cities"[43] define as smart a city connecting the physical infrastructure, the IT infrastructure, the social infrastructure, and the business infrastructure to leverage the collective intelligence of the city.

In this respect, SCOs constitute an important development that is expected to have a profound impact on the socioeconomic future of Europe. ICT is a strong enabler for cities to turn "smarter" and thus offer their citizens the opportunity of a better quality of life. This can be achieved through better decision-making about a variety of domains within a city.

Particular areas where SCOs find fertile ground for development include, among others, energy consumption minimization, the organization of light and traffic, management of public as well as private transport, and systems related to environmental protection and health care.

Those indicative areas are outlined in Fig. 3.12.

3.3.2 *SCOs Challenges*

Smart cities worldwide are becoming increasingly smarter, through capitalizing on new technologies and insights to transform their systems and operations delivery to citizen-centered useful service delivery.[44] To be able to continue advancing in this area and consolidate a solid "smart background", several fundamental requirements need to be addressed from an operational point of view.[45] To extract those requirements, a set of smart city operations challenges, as identified in the international literature, are detailed below.

3.3.2.1 Level of Intelligence ("Smartness") Required

Intelligence ("smartness") might be a difficult concept to sketch from various viewpoints. As such, a city should appropriately consider a priori the desired levels of smartness to be achieved at short, medium, and long time scale. This depends of course to a number of services that a city wants to provide to its citizens so as to be considered "smart." Moreover, to do so, a city should consider the needs, plans, and

(SRF), Vienna University of Technology. Available from http://www.smartcities.eu/download/smart_cities_final_report.pdf

[43] Harrison, C., Eckman, B., Hamilton, R., Hartswick, P., Kalagnanam, J., Paraszczak, J., & Williams, P., "Foundations for Smarter Cities", IBM Journal of Research and Development, 54(4)

[44] Toppeta, D., The Smart City Vision: How Innovation and ICT Can Build Smart,"Livable", Sustainable Cities, 2010, The Innovation Knowledge Founda-tion. Available from http://www.thinkinnovation. org/file/research/23/en/Toppeta_Report_005_2010.pdf

[45] http://www.hitachi.com/products/smartcity/download/pdf/whitepaper.pdf, accessed March 16th, 2015

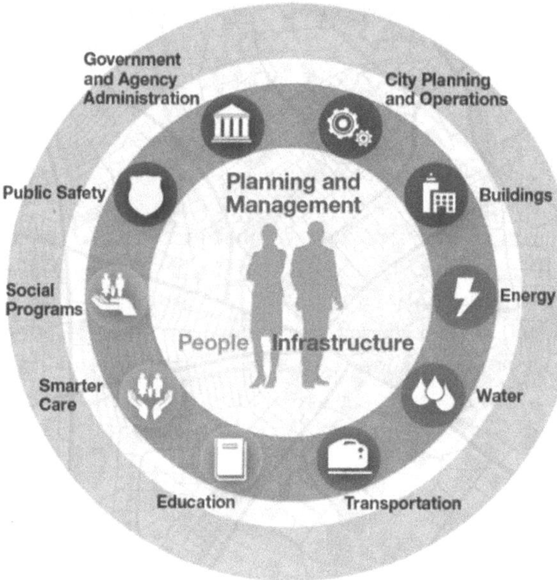

Fig. 3.12 Fundamental SCO. http://www.ibm.com/smarterplanet/us/en/smarter_cities/overview/, accessed February 26th, 2015

opinions of all stakeholders involved in its operations, such as (1) citizens, (2) service providers, (3) businesses, (4) municipal authorities, and (5) national standards. At the same time, all economic, environmental, and people-oriented viewpoints should be considered. Last, scalability of the smart operations to be provided should also be considered. This means achieving a balance not just between the interests of a particular city's stakeholders, but also taking into account relationships with neighboring cities. The above seems a complex algorithmic process with multiple variables.[46,47,48]

3.3.2.2 Technology

Undoubtedly technology constitutes the primary driver towards the transformation of a city from a conventional one to a smart one. A smart city usually is based on a set of technologies, comprising mainly smart computing ones, that operate in relation to critical infrastructure services and components. These smart computing

---

[46] Difallah, D.E., Cudre-Mauroux, P., McKenna, S.A., "Scalable Anomaly Detection for Smart City Infrastructure Networks", Internet Computing, IEEE, Volume:17, Issue: 6, 2013, pps. 39–47

[47] Giffinger, R., Fertner, C., Kramar, H., Kalasek, R., Pichler-Milanović, N., & Meijers, E., "Smart Cities: Ranking of European Medium-Sized Cities", Vienna, Austria: Cen-tre of Regional Science (SRF), Vienna University of Technology. Available from http://www.smartcities.eu/download/smart_cities_final_report.pdf

[48] Harrison, C., Eckman, B., Hamilton, R., Hartswick, P., Kalagnanam, J., Paraszczak, J., & Williams, P., "Foundations for Smarter Cities", IBM Journal of Research and Development, 54(4)

technologies are "new generation" technologies, comprising software, hardware, and networks. Their purpose is to enhance current IT systems in order for them to obtain (1) real-time awareness of the world and (2) advanced analytics capabilities in order to assist citizens towards more intelligent decisions in a variety of fields.[49]

In this respect, numerous challenges can be identified, such as (1) people (citizens, employees, etc.) that have limited integration skills, (2) missing interdepartmental coordination in the management of public and private entities, (3) limited cross-sectoral cooperation, and others.

3.3.2.3 Scalability of Smart Solutions

Smartness should be scalable enough, in that a city should appropriately design the objectives to be achieved at various scales.

First comes the minimum objectives that will attribute "smart" characteristics to the city and will be able to provide its citizens the minimum levels of quality needed to live a civilized life. At this high-level stage, the values of a city and its residents include many qualitative concepts and things of an emotional nature, such as lifestyle values and a sense of attachment to the neighborhood.

Second comes the fundamental objectives that will enhance the level of smartness of the city towards a desired level, such as the reduction of carbon emissions. Such objectives could be agreeable at a local/regional/national level.

Last comes some longer term objectives that will further advance the smartness level already achieved, which are usually set at a local level, albeit being negotiable also at an international level in the context of organizations and fora.

3.3.2.4 Formulation of City-Specific Objectives

A city usually sets at a local level some standards to be achieved at various time scales. Then, some Key Performance Indicators (KPIs) are monitored so as to evaluate the achievement of those standards. Those KPIs are nothing less than city-selected criteria/benchmarks. Moreover, KPIs should be adaptive enough to respond to new (external) requisitions.[50,51,52]

---

[49] Washburn, D., Sindhu, U., Balaouras, S., Dines, R. A., Hayes, N. M., & Nelson, L. E., "Helping CIOs Understand "Smart City" Initiatives: Defining the Smart City, Its Drivers, and the Role of the CIO. Cambridge, MA: Forrester Research, Inc. Available from http://public.dhe.ibm.com/partnerworld/pub/smb/smarterplanet/forr_help_cios_und_smart_city_initiatives.pdf

[50] Naphade, M., Banavar, G., Harrison, C., Paraszczak, J., Morris, R., "Smarter Cities and Their Innovation Challenges", IEEE Computer, Volume: 44, Issue: 6, 2011, pps. 32–39

[51] Hogan, J., Meegan, J., Parmar, R., Narayan, V., Schloss, R.J., "Using stan-dards to enable the transformation to smarter cities", IBM Journal of Research and Development, Volume: 55, Issue: 1.2, 2011, pps. 4:1–4:10

[52] Difallah, D.E., Cudre-Mauroux, P., McKenna, S.A., "Scalable Anomaly Detection for Smart City Infrastructure Networks", Internet Computing, IEEE, Volume:17, Issue: 6, 2013, pps. 39–47

In order to formulate city-specific objectives, factors that need to be taken into account are (1) the people and cultural diversity and (2) the environment.[53,54]

3.3.2.5 Economic Growth

From a high-level, economics viewpoint, a city can be thought of as an entity that enables internally operating business groups to obtain income from outside its geographical region, and then enables the obtained revenues to circulate within its region. This of course can function the other way round (extroversion).

Accordingly, the economic performance of a city can be viewed from two viewpoints: its industrial competitiveness relative to other regions and the soundness of the finances within its region.

In this respect, it is essential that when planning and designing the provision of smart city operations, one must take a holistic, long-term approach. In particular, the assessment of strengths, weaknesses, opportunities, and threats needs to look 10 or even 20 years ahead. Such a process will allow a city to continue attracting immense attention for businesses, whilst being comfortable and secure for its citizens.[55,56]

3.3.2.6 Management and Organization

The effective management and organization of a city is highly related to smart city initiatives; nevertheless, only a few studies in the academic literature adhere to address such issues. On the other hand, several research efforts on IT initiatives and projects have identified these initiatives as critical success factors and challenges.[57,58] To that end, aspects regarding management and organization in smart city initiatives have to be investigated, focusing on the thorough analysis of e-government and IT projects.

---

[53] Benouaret, K., Valliyur-Ramalingam, R., Charoy, F., "CrowdSC: Building Smart Cit-ies with Large-Scale Citizen Participation", Internet Computing, IEEE, Volume: 17, Issue: 6, 2013, Page(s): 57–63

[54] Walravens, N., Ballon, P., "Platform business models for smart cities: from control and value to governance and public value", Communications Magazine, IEEE, Volume: 51, Issue: 6, 2013, pps. 72–79

[55] Naphade, M., Banavar, G., Harrison, C., Paraszczak, J., Morris, R., "Smarter Cities and Their Innovation Challenges", IEEE Computer, Volume:44, Issue: 6, 2011, pps. 32–39

[56] Hogan, J., Meegan, J., Parmar, R., Narayan, V., Schloss, R.J., "Using standards to enable the transformation to smarter cities", IBM Journal of Research and Development, Volume: 55, Issue: 1.2, 2011, pps. 4:1–4:10

[57] Gil-García, J. R., & Pardo, T. A., "E-government success factors: Mapping practical tools to theoretical foundations", Government Information Quarterly, 22(2), 187–216.

[58] Scholl, H. J., Barzilai-Nahon, K., Ahn, J-H., Olga, P., & Barbara, R., "E-commerce and e-government: How do they compare? What can they learn from each other?", in Proc. 42nd Hawaiian International Conference on System Sciences (HICSS 2009), Koloa, Hawaii, January 4–7.

In this respect, the authors of [58] suggested several challenges, namely: (1) the size of the project, (2) behaviors and attributes of managers, (3) limited alignment of the organizational goals, (4) having numerous goals that can be conflicting, and (5) cases of having resistance to change.

3.4 Sustainable Mobility: Mobility as a Service (MaaS)

The latest mass transit and e-mobility technologies match with city infrastructures from monorail and metro systems running through buildings at-grade, elevated or underground, to new solutions for electric vehicles. These solutions support a better way, which helps us thinking from traditional transport modes to electric public transport.

Smart mobility is a key challenge in the world. The huge increase in urban population and the growing environmental topic find prosper ground to the concept of smart mobility, which proposes solutions for greener, safer, and more efficient transfer of humans and freight.

It can be stated that mobility has been seen throughout the years as a product. That "mobility product" comprises physical infrastructure, vehicles, as well as fuels which used people to mobilize. But, mobility is approached as a service also. This means that mobility is a means to achieve a number of goals, such as engaging in economic activity, providing food, and having access to entertainment activities. Mobility is also related to the usage of mobile phones; we use mobile applications to manage our lives "on the go." All these recently introduced capabilities are based on infrastructure (either physical or digital) with significant potential. When we supplement management practices for cities with digital technologies, it is possible to improve the mobility services provided to citizens, and at the same time manage the demand on physical transport networks and generate wider environmental sensitivity.[59]

In this way, the challenges in smart mobility are[60]:

1. To develop a system that can communicate with the vehicle and so the user is capable of receiving information from the environment (context information), which can have influence in the overall performance of the vehicle. That information may comprise traffic information, internet-connected vehicles, parking management, carpooling, etc.
2. To make the best effective use of the trip planning and routing of fully electric vehicles, exploiting data from these sources including alternatives from other transport modes adapted to user's needs.
3. To set effective and optimal charging strategies which match to user and fully electric vehicles needs and grid conditions.

---

[59] Charbel Aoun, "Urban mobility in the smart city age", 2012

[60] "Smart mobility in smart city". Available from http://www.mobincity.eu/about_mobincity/project_overview

4. Using energy-saving methods (as driving modes and In-Car Energy Management Services) within the fully electric vehicles interaction with the driver.

An example is a new mobility model, the Mobility-as-a-Service (MaaS). MaaS bridges the gap between public and private transport operators, envisaging the integration of all the fragmented tools (planning, booking, real-time information, payment, and ticketing) a traveler needs to conduct a trip. This model reduces the dependence on private vehicles and allows modern travelers in urban areas to plan and manage their transit quickly and safely using their smartphones. The key to successful uptake of such services is the effective integration among different technologies and tools.

Several EU funded initiatives are relevant to the smart mobility in smart cities in general, as well as MaaS in particular. The InSMART[61] concept brings together cities, scientific and industrial organizations in order to establish and implement a comprehensive methodology for enhancing sustainable planning addressing the current and future city energy needs through an integrative and multidisciplinary planning approach. READY4SmartCities[62] operates in a European context where other initiatives are currently running in order to create a common approach on Smart Cities. STEP-UP—Strategies Towards Energy Performance and Urban Planning[63] takes an integrated approach to energy planning, integrated project design, and implementation by addressing 3 vital themes together: energy and technology; economics; and organization and stakeholders.

TRANSFORMation Agenda for Low Carbon Cities[64] supports cities to meet the 20-20-20 targets by the integration of energy in urban management. PLEEC Planning for energy efficient cities[65] gathers cities with innovative planning and ambitious energy-saving goals. SMART-ACTION[66] supports the development of strategic research agendas and serves as an enabler for the dissemination and further integration of results into future research and industrial developments, while coordinating international efforts.

3.5 Case Studies

3.5.1 Traffic Assessment, Forecasting, and Management Applications (TAFM)

Nowadays, TAFM applications constitute the information and communication systems behind modern transportation companies. They allow a communication between the control center and the vehicles of a transportation company, a

[61] http://cordis.europa.eu/project/rcn/186975_en.html, accessed December 2015
[62] http://cordis.europa.eu/project/rcn/110042_en.html, accessed December 2015
[63] http://cordis.europa.eu/project/rcn/186983_en.html, accessed December 2015
[64] http://cordis.europa.eu/project/rcn/186978_en.html, accessed December 2015
[65] http://cordis.europa.eu/project/rcn/186984_en.html, accessed December 2015
[66] http://cordis.europa.eu/project/rcn/109708_en.html, accessed December 2015

computer-controlled transport operation, transfer synchronization between different transportation companies and provide dynamic passenger information.

The current changes from the originally isolated system for public transport to an Intermodal Transportation Control System is not only pushed by a rapid evolution of ICT but also politically intended. In the EU directive to accelerate ITS adoption, governmental support is a fundamental prerequisite. Major goals are the seamless traffic and passenger information in real time and the integration of intermodal transport for a seamless traffic management.

Many small- and medium-sized transportation companies are now challenged to invest in new or partly renew their existing control systems under increasingly difficult financial conditions. Here cost-effective, secure solutions, adaptable to the needs of each transportation company, are required.

Due to existing challenges like a better scalability of standard TAFM as soon as the reduction of acquisition and operational costs the so-called TAFM Light systems were created. Due to their reduced functionality (often limited to passenger information and transfer synchronization) these systems are affordable alternatives to classical TAFM for transportation companies and associations. But the enlargement with new functionality is linked with enhanced effort.

Also the shared use of a TAFM by several transportation companies aims to a flexible access to the functionality of a TAFM at simultaneous use of savings potentials. As seen by many actors of public transport, these are the first steps to an improved scalability.

3.5.2 Road Luminosity Management Applications

Those kinds of services will generate and visualize knowledge regarding the best levels of luminosity for a given area, for a specific time span, having in mind the trends in terms of mobility issues and the number of citizens in the area, the respective luminosity standards and any feedback from the users. Based on this knowledge, usually such tools will suggest the luminosity value that should be applied at the future period of interest. The suggested value will also be accompanied by an accuracy metric. Therefore, they serve as forecasting tools for the future luminosity needs. Moreover, the deviation of the current luminosity values (deriving from the current capabilities of the infrastructure) from the predicted optimal ones will also be provided. The outcome of this service can be depicted in a user-friendly map of the area which will depict:

(a) The proposed value for luminosity for a specific period of time and a specific area, as well as an accuracy metric related to the suggestion.
(b) The level at which current luminosity values may vary from the ones suggested.

3.5.3 *Car Pooling (Ride-Sharing)*

Car pooling is an idea that has been proposed by several researchers since many years, mainly for reducing traffic. However, with the advent of ICT, this idea has turned into a smart mobility service offered by some cities/regions to their citizens and visitors.

The service assumes that we have a set of drivers that form the drivers' pool, as well as a set of passengers, that form the passengers' pool. There is an association between these two pools, which is based on context information parameters. That information may include for instance data on the passengers' and drivers' current positions and itineraries. Moreover, these two pools are related with parameters regarding personal profiles and services. Last, a set of overarching policies reflects driver/passenger preferences, in the form of weights (importance) attributed to the aforementioned parameters.

3.5.3.1 Business Case

As depicted in Fig. 3.13, we have the two pools already defined: the one with the drivers and the other one with the passengers. As already stated, there is an association between these two pools, which is based on context information parameters (for instance, data on their current position) and this is shown in Fig. 3.14a. Furthermore, the two pools are related to personal profile parameters (an indicative set of which is depicted in Fig. 3.14b), as well as with service-related parameters (shown in Fig. 3.14c).

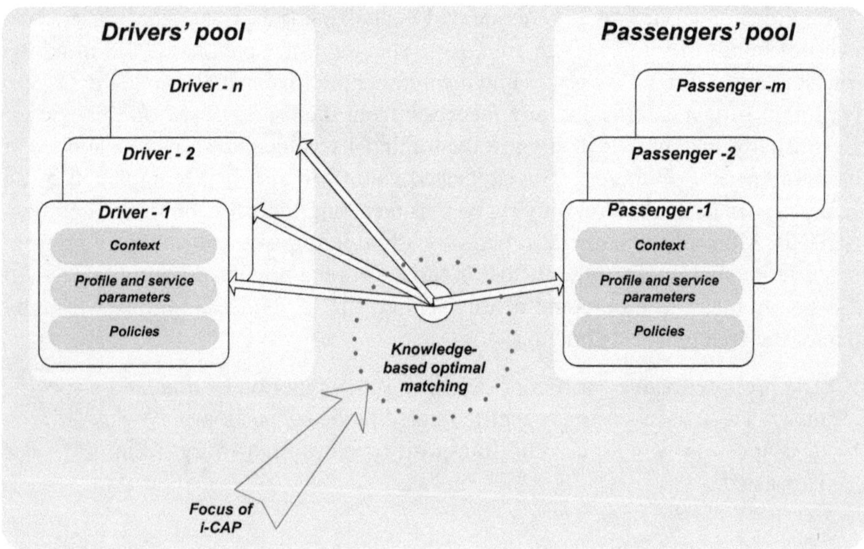

Fig. 3.13 i-CAP context of operation

(a)

Driver Parameters	Passenger Parameters
Driver departure point	Passenger departure point
Driver departure time	Passenger departure time
Driver destination point	Passenger destination point

(b)

Personal profile parameters
1. Social behavior
2. Driving skills
3. Driving style
4. Repeat willingness
5. Age
6. Smoking
7. Gender
8. Educational level

(c)

Service parameters
1. Safety of ride
2. Punctuality
3. Velocity
4. Itinerary cost
5. Primary road usage

Fig. 3.14 (**a**) Context information, (**b**) personal profile parameters, (**c**) service parameters

Finally, a set of overarching policies reflects driver/passenger preferences, in the form of weights (importance) attributed to the aforementioned parameters.

In regard to these parameters, at least the personal profiles and the services are subject to changes through time. In that framework, the main target of the functionality is to interact with the pool of drivers (for each one of the passengers) in order to indicate an optimal match. This indication/decision will be based on the request, the context, and any available data related to personal and service profiles, as well as previously obtained knowledge that has now been turned into experience. An ICT-based management functionality can guarantee the communication; such a functionality is the one proposed herein: i-CAP.

According to the specific business case, we have a passenger/user that aims to carry out a specific journey. The user then logs on to i-CAP which may be installed in a mobile phone or laptop and at the same time form part of the communication infrastructure: i-CAP can facilitate the required exchange of data using a web-based interface. If the passenger uses i-CAP for the first time, he is guided to fill in a form related to any possible preferences he/she may have. A number of virtual tests are then carried out by i-CAP so that to identify the best matches for our user, and therefore being able to converge faster when the time comes. If the user has already registered to i-CAP, he can proceed with making a request immediately, as the system is able to recognize the specific user and obtain access to any required personal preferences and information as well as historical data. All that data (including personal preferences and information as well past activity/history) is maintained in log files which are stored in appropriate databases. Apart from the users/passengers, the system is aware of all members of the drivers' pool that are available, and thus it is capable of finding the best matches between passengers and drivers.

Another possibility is to have the aforementioned scenario initiated by the driver, with a quite similar evolution. At any case, we have four basic requirements for i-CAP:

1. *Personalization*. This requirement is essential to verify that the system provides valid matching solutions, best fitting the needs of both passenger and the driver.
2. *Adaptability*. It is critical to ensure effective interaction of the system with the drivers as well as that the system efficiently exploits the preferences of passengers.
3. *Knowledge aggregation*, which is critical in order to utilize information that is gathered from previous interactions, so that to reduce delays in upcoming decisions and make them more efficient.
4. *Scalability*, in order to ensure that the system is able to select in an ad hoc manner, based on the particular contextual needs, either a more distributed or a more centralized mode of operation.

In the light of those requirements, the proposed functionality (i-CAP) is shown in Fig. 3.15.

As shown in the figure, the system's input include (1) information regarding the context, (2) data and parameters related to personal profiles and services, and (3) information on policies that are responsible to assign weight to the different parameters. The output of the system is the best possible "driver–passenger" matching. The overall process consists of specific phases: the "robust discovery phase" and the "decision-making phase." The first one focuses on the exploitation of a Bayesian-based model towards the maximization of the probabilities that the parameters will reach certain values. The specific model assists the system to obtain knowledge

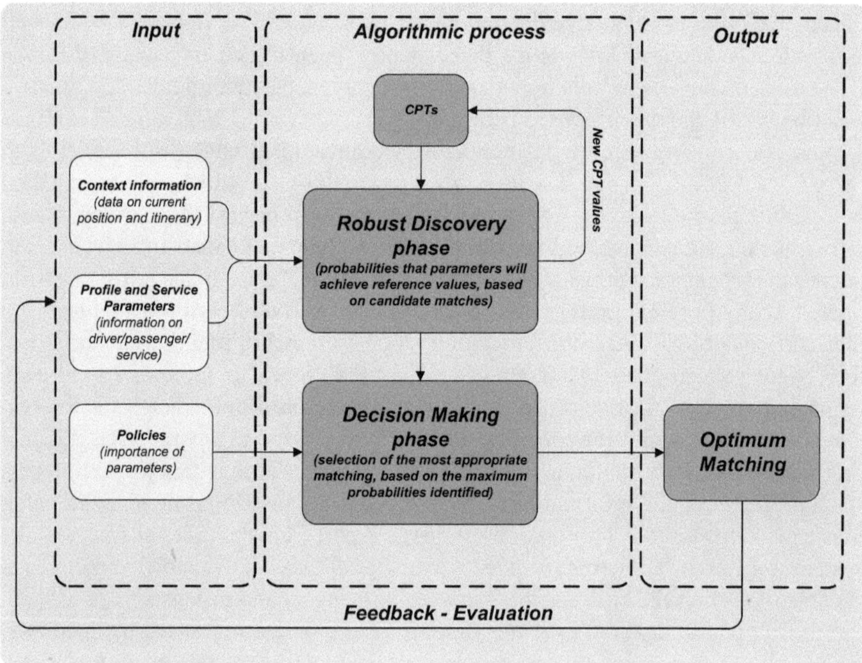

Fig. 3.15 i-CAP functionality description

gradually. The second phase on the other hand gets these probabilities, and based on them it estimates the best possible matching.

The gathering of knowledge from the system is empowered by an evaluation procedure. More specifically, passengers are to evaluate the drivers after the ride sharing is completed. The evaluation is carried out based on specific parameters using a ten point integers' scale, in the form of utility volumes. In this scale, lower values stand for evaluations closer to "poor" while larger values stand for evaluations closer to "excellent." Utility volumes express the level of satisfaction of each carpooler. Ranking might concern all available parameters and serves as an input to the Bayesian-based model.

3.5.3.2 Formal Description

Input

The following analysis focuses on a prospective passenger willing to organize a journey. As already discussed, the input to the management system comprises context information, data related to personal and service profile, and information related to policies. These concepts, while they are quite general, they lead to specific data structures that are summarized in Table 3.1. These are the *candidate drivers*, the passengers'/drivers'/service's *parameters*, and the *importance* of each parameter dictated by passengers'/drivers' profiles.

Let us define the set of potential passengers, *PP*. We use *P* in order to represent a passenger. This means that *P* may take values varying from 1 to |*PP*|. Similarly, let

Table 3.1 Data structures involved in the i-CAP functionality

D, V_j	Random variables representing the candidate driver and the value of the *j*-th parameter ($j = 1,\ldots,N$)
RV_{ij}	Set of reference values that can be taken by random variable V_j when driver $i \in CD$ is considered
dif_{ij}, rv_{ij}^k	Difference between the larger and the lower reference value of RV_{ij}, and *k*-th reference value for parameter *j* when driver *i* is considered
$X_i = RV_{i1} \times \cdots \times RV_{iN}$	Set of vectors of reference values for candidate driver *i*
$f\left(\bar{x},i\right)$	Probability of each vector $\bar{x} \in X_i$
$\Pr\left[V_j = rv_{ij}^k \mid D = i\right]$	Conditional probability that parameter *j* can achieve the *k*–th reference value, given that driver *i* is considered
rv_{ij}^{coll}, cor_{ij}^k, nf_{ij}	Most recent value collected, achieved by driver *i* regarding parameter *j*, correction in the conditional probability $\Pr\left[V_j = rv_{ij}^k \mid D = i\right]$, and normalization factor
$pr_{\max}$	Maximum value allowed for the conditional probabilities
OF_i	Objective function value for driver *i*

us define the set of potential drivers, *CD*. In the same way we use *D* to represent the driver. This means that *D* can take values varying from 1 to $|CD|$.

Having defined passengers and drivers, we denote as *N* the set of parameters. Each one of these parameters, j $(j = 1,\ldots,N)$, may refer to a specific thing (such as, for instance, gender, age, gender, driving style, and others) and has a specific weight expressing its importance, which is denoted with w_j, where $\sum w = 1$. The w_j values define a vector of weights w.

As already stated, the values of the parameter may change through time randomly. To that end, random variables i and v_j $(j = 1,\ldots,N)$ are used in order to represent the driver and the value of the j-th parameter, respectively. Each variable V_j is related with a set of reference values RV_{ij} $(i \in CD)$. Variable v_j can take a value of those in RV_{ij}, when driver i is considered.

The system needs to develop knowledge which is based on conditional probabilities, that can be expressed with the form $\Pr\left[V_j = rv_{ij}^k \mid D = i\right]$, where $rv_{ij}^k \in RV_{ij}$ expresses the k-th reference value for the j-th parameter when driver i is considered.

For each driver $i \in CD$, a set X_i can be defined. The members of set X_i derive from the Cartesian product of the RV_{ij} ($j = 1,\ldots,N$) sets, i.e., $X_i = RV_{i1} \times RV_{i2} \times \ldots \times RV_{iN}$. Therefore, a member $\tilde{x}$ of X_i has the form $\tilde{x} = \left\{rv_{i1}^{k1},\ldots,rv_{iN}^{kN}\right\}$, where $rv_{ij}^{kj} \in RV_{ij}$ $(j = 1,\ldots,N)$ and k_j $(j = 1,\ldots,N)$ are integers.

Probability density function. Based on the previous definitions, the following probability density function can be defined:

$$f\left(\tilde{x},i\right) = \Pr\left[V_1 = rv_{i1}^{k1},\ldots,V_N = rv_{iN}^{kN}, D = i\right] = \Pr\left[D = i\right]\cdot\prod_{j=1}^{N}\Pr\left[V_j = rv_{ij}^{kj} \mid D = i\right] \tag{3.1}$$

where $i \in CD$, $\tilde{x} \in X_i$, $rv_{ij}^{kj} \in RV_{ij}$ ($j = 1,\ldots N$), and k_j ($j = 1,\ldots N$) are integers.

The $\Pr\left[D = i\right]$ probabilities express the level of information that is available for each driver i. The sum of the $\Pr\left[D = i\right]$ quantities, over the drivers $i \in CD$, is 1. As expected, as much information we have for a specific driver, the more reliable is the knowledge of the system regarding that driver, and hence the higher the $f\left(\tilde{x},i\right)$ values.

This means that the values of the $f\left(\tilde{x},i\right)$ function depict in a cumulative way our knowledge on how possible is to have a specific value for a parameter indicated in $\tilde{x}$, by driver i. The sum of the $f\left(\tilde{x},i\right)$ values, overall $\tilde{x} \in X_i$ and $i \in CD$, is one. In other words, the $f\left(\tilde{x},i\right)$ value depicts the probability of the ($\tilde{x}$, i) pair, with respect to all others possible pairs. To that end, using this function we can come up with the higher reliability associated with the selection of a specific driver.

Objective and Solution

The overall purpose is to choose the most appropriate driver from the ones available in *CD*. Let us now describe in more details the two phases already mentioned, in order to achieve that objective: the robust discovery phase and the decision-making phase.

Robust discovery phase. The target of this phase is to estimate the most probable values for our parameters. In order to achieve that, the probabilities in the right end of (3.1) need to be updated, and this is why i-CAP gathers evaluation data that has been provided for the *CD* drivers. Updating the probabilities in (3.1) can take into account the "distance" of the collected evaluation values from the reference values. For instance, suppose that according to the most recent evaluation, driver i can achieve rv_{ij}^{coll} regarding parameter j. Let us use dif_{ij} in order to express the difference between the maximum and the minimum reference value in RV_{ij}. For each reference value, $rv_{ij}^{k} \in RV_{ij}$, we can have a correction factor estimated by:

$$cor_{ij}^{k} = 1 - \left(\left| rv_{ij}^{k} \quad rv_{ij}^{coll} \right| / \, dif_{ij} \right) \tag{3.2}$$

As $0 \le cor_{ij}^{k} \le 1$, if we have a value closer to one, then the reference value is close to the collected one, meaning that the corresponding conditional probability value should be reinforced accordingly. This is not the case though, if cor_{ij}^{k} is closer to zero.

In order to obtain the updated conditional probabilities we use the following expression:

$$\Pr\left[V_j = rv_{ij}^{k} \middle| D = i \right]_{new} = nf_{ij} \cdot cor_{ij}^{k} \cdot \Pr\left[V_j = rv_{ij}^{k} \middle| D = i \right]_{old} \tag{3.3}$$

Parameter nf_{ij} in the aforementioned equation is a normalizing factor used in order to guarantee that all the updated probabilities will sum up to one. In addition to that, and in order to verify adaptability to new conditions, we have an upper limit for the conditional probabilities, $pr_{\max}$.

Summing up the update strategy, it comprises four main parts, namely: (1) collecting the parameter reference values based on evaluations; (2) computing the correction factors using (3.2), as well as the new probabilities using (3.3); (3) verify that in case a probability is over $pr_{\max}$ then it is set equal to this upper threshold, and (4) calculation the new normalizing factors by forcing the remaining probabilities to sum to $(1 - pr_{\max})$, and the new values are computed for the remaining probabilities.

Decision-making phase—exploitation of knowledge. The system aims to select the drivers that have high probability of achieving the most suitable parameter values. In order to model these aspects an Objective Function (OF) value, OF_i, is defined for each driver, $i \in CD$. In order to estimate these OF_i values for each one of the drivers, we use the following equation

$$OF_i = \sum_j \left\{ \max\left(\Pr\left[V_j = rv_{ij}^{k} \mid D = i \right] \right) \right\} \cdot w_j \tag{3.4}$$

where $i \in CD$, ($j = 1,\ldots,N$) and $rv_{ij}^{k} \in RV_{ij}$ expresses the *k*-th reference value for the *j*-the parameter and for the driver *i*.

The driver with the highest OF_i value should be selected based on the knowledge as obtained from the aforementioned process.

3.5.3.3 Results

This section describes the behavior of the proposed system for both phases already described. We investigate specific aspects such as the evolution of conditional probabilities, probability density function values, and driver OF values.

In order to design the following scenarios several research efforts have been used.[67,68,69,70] They derive from the inputs of the functionality, that include as already stated information related to the context, the personal and service profiles, and the policies, which lead to the set of available drivers, the values of the available parameters and the importance of these parameters, expressed through their weights. The target of these scenarios is to verify how fast this system can converge to a specific solution during the robust discovery phase, as well as to estimate the best possible matching during the decision-making phase.

To that end, we have three specific scenarios. Scenario 1 is a regular carpooling scenario, with the aim to show the gradual development of knowledge while the system tries to identify, as fast as possible, the most likely parameter values, leading to the most suitable matching/decision. Scenario 2 focuses on how fast the system adapts when one specific parameter has significantly increased importance with respect to the others. Finally, Scenario 3 examines the impact on the decision-making process of the case where a driver receives continuous positive evaluations from passengers and thus improves its "reputation."

Scenario 1: Regular Service

This scenario focuses on proving how efficient i-CAP can be in a regular service case, i.e., the case where five personal profile and service parameters are considered, each one with its own importance (weight), as shown in Fig. 3.16a. Four reference values are considered for each parameter (also shown in Fig. 3.16a).

---

[67] C. Steger-Vonmetz. Improving modal choice and transport efficiency with the virtual ridesharing agency. Proceedings of the 8th International, IEEE Conference on Intelligent Transportation Systems, 2005

[68] F. Wang, C. Herget, D. Zeng, "Developing and Improving Transportation Systems: The Structure and Operation of IEEE Intelligent Transportation Systems Society", IEEE Transactions on Intelligent Transportation Systems, Volume 6, Issue 3, Sept. 2005 Page(s):261–264

[69] Darm—Division of Resources Management. Carpooling and You. Public domain document, Florida, U.S., January 2005. Available online: http://www.dep.state.fl.us/Air/publications/airpubs/carpool.pdf

[70] R. Nolanda, W. Cowartb, L. Fultonc, "Travel demand policies for saving oil during a supply emergency", Energy Policy Journal. Volume 34, Issue 17 pp. 2994–3005 Elsevier 2006

(a)

Parameters	Reference values				Weight
Driving skills	1	4	7	10	$W_{drivingskills} = \mathbf{0,2}$
Safety	1	4	7	10	$W_{safety} = \mathbf{0,3}$
Primary roads usage	1	4	7	10	$W_{primaryroads} = \mathbf{0,1}$
Itinerary cost	1	4	7	10	$W_{cost} = \mathbf{0,2}$
Social behavior	1	4	7	10	$W_{socialbehavior} = \mathbf{0,2}$

(b)

Parameters	$D = 1$	$D = 2$	$D = 3$
Driving skills	7	9	7
Safety	8	8	6
Primary roads usage	9	7	10
Itinerary cost	6	10	8
Social behavior	10	6	9

Fig. 3.16 Scenario 1—(**a**) parameters and respective weights, (**b**) uniform distribution of parameter values collected through the evaluation procedure, for the 3 drivers

It is important to note here that more reference values are expected to lead to better results. In addition to that, the distance between two subsequent reference values is not necessarily the same. At the beginning, the conditional probabilities ($\Pr\left[V_j = rv_{ij}^k \mid D = i\right]$), $j = 1\ldots5$, are uniformly distributed, i.e., equal to 0.25 for the 4 reference values of each parameter. This is due to the fact that initially there is no previous knowledge available for the system. According to this specific scenario, we have three possible drivers, with $\Pr\left[D = i\right] = 0.33$, $i = 1\ldots3$. Figure 3.16b depicts the values of the parameters collected through the evaluation procedure. These values are extracted with the addition of all values and the division of the sum by the number of evaluations. Last, the parameter values collected are supposed to be uniformly distributed in the parameter values space (we assume a space of "6"–"10" for facilitating the calculations; however, this can be completely "ad hoc" amended) for increasing the reliability of the functionality's operation.

Robust Discovery Phase

This phase focuses on how the conditional probabilities evolve and the values of the probability density function that express the level of knowledge available in the system. As presented, the conditional probabilities and the probability density function values constitute the basis, on which knowledge can be built.

The algorithm is applied in 15 series of runs (computations). By using (3.2), we calculate the correction factors. Then, by using (3.3), we compute the new (adjusted) conditional probabilities.

Figure 3.17a–c presents the information for the first driver: it includes the conditional probabilities' distribution and the three sub-figures refer to each one of the three different parameters (safety, cost, and driving skills). In each sub-figure of Fig. 3.17, there is no knowledge for the system at the beginning, but we can see that i-CAP readily learns the capabilities of the parameter and converges to the conditions indicated by the evaluations. These remarks are backed up by the fact that regarding, e.g., the "safety" parameter, the conditional probability $\Pr\left[V_{safety}=7\mid D=1\right]$ immediately becomes significant and, very soon, reaches levels that are close to the highest possible ones. As expected in this case (see also Fig. 3.17a), from the beginning there are high values also for $\Pr\left[V_{safety}=10\mid D=1\right]$, a slight diminishment for $\Pr\left[V_{safety}=4\mid D=1\right]$, and a severe degradation for $\Pr\left[V_{safety}=1\mid D=1\right]$.

Following the same approach, as we move forward to the next iterations, the most probable value is further reinforced (in the specific scenario, this is the value 7). Similar lines can be extracted for the other drivers. In every case, the system adapts fast and effectively to the desired parameter values.

If the conditional probabilities change, this means that the same happens with the probability density function in (3.1)—and as a result, our knowledge regarding the capabilities of the drivers also undergo changes. The probability density function values for the 3 drivers, i.e., $f\left(\tilde{x},i\right), i\in CD$, are shown in Fig. 3.18a. From the figure we may derive that gradually gathering knowledge seems to be easier for the first driver, as it comes to a significant level after only a few iterations. On the other hand, in the case of the third driver knowledge gathering seems to be more difficult and to need more time. In the case of the second driver we have knowledge gathered somewhere in between in terms of delays. But we can conclude that the reliability of the decisions that the system has to take increases within a small number of iterations.

Decision-Making Phase

The first phase of the whole process we saw that in the case of the first driver the reliability level is greater. This is depicted in the second phase, as the system actually "chooses" that first driver for a match. The OF values for each driver are calculated using (3.4) and shown in Fig. 3.18b. The system decides that the first driver is the best choice, taking into account context, personal and service profile, as well as policies information.

In more detail, the system needs five iterations to identify that the 1st driver is best choice, which is assumed as a very good result, as it can catch improvements in the potential behavior of a driver that are only temporal. If such improvements were accepted it could lead to frequent changes, and potential oscillations, in the selected drivers. The small number of steps results in more fast adaptations.

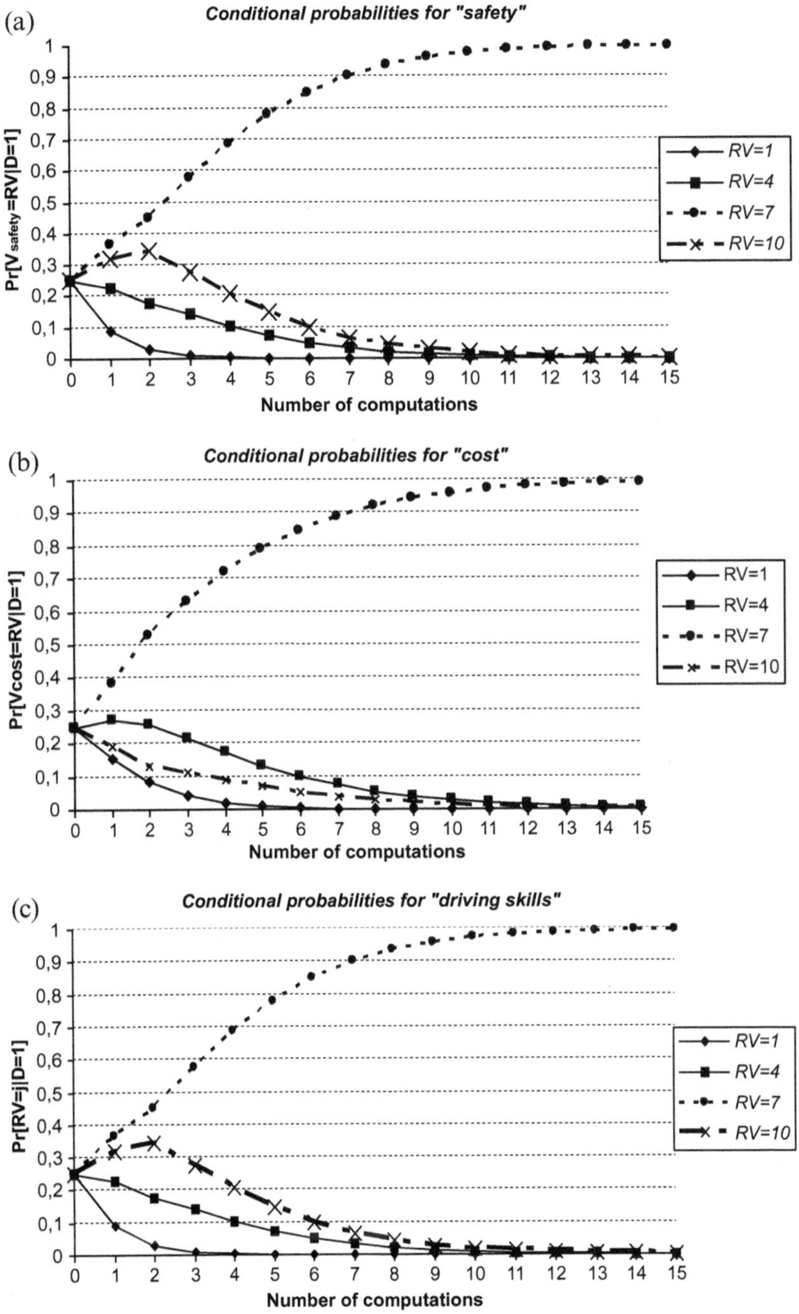

Fig. 3.17 Scenario 1, first driver—(**a**) conditional probabilities for parameter "safety," (**b**) conditional probabilities for parameter "cost," (**c**) conditional probabilities for parameter "driving skills"

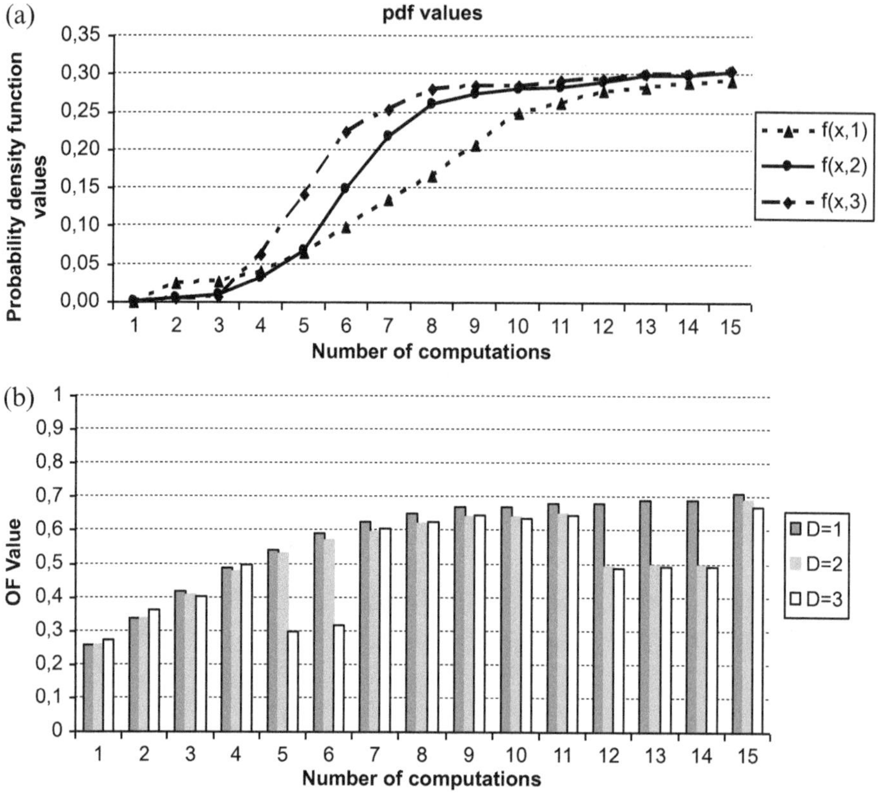

Fig. 3.18 Scenario 1—(**a**) probability density function values for the 3 drivers ($f\left(\bar{x},i\right)$), (**b**) OF values of the 3 drivers

Scenario 2: Cost-Driven Scenario

This scenario focuses on the evaluation of the performance of i-CAP when one specific parameter (the cost, in our specific scenario) has a very high value. All the details regarding the parameters and their values are depicted in Fig. 3.19a, while the values collected through the evaluation process can be found in Fig. 3.19b.

Robust Discovery Phase

As in the previous scenario, we have 15 iterations of computations. Figure 3.20a–c shows the distribution of conditional probabilities for the case of the third driver ($D=3$) and the parameters also examined in the previous case. As shown, the system learns very fast the capabilities of the parameter and converges to the value indicated by the evaluation process. Focusing on the parameter of "cost," the conditional probability, $\Pr\left[V_{cost}=10 \mid D=3\right]$, immediately becomes dominant, in

Parameters	Reference values				Weight
Driving skills	1	4	7	10	$W_{drivingskills} = \mathbf{0{,}2}$
Safety	1	4	7	10	$W_{safety} = \mathbf{0{,}07}$
Primary roads usage	1	4	7	10	$W_{primaryroads} = \mathbf{0{,}03}$
Itinerary cost	1	4	7	10	$W_{cost} = \mathbf{0{,}6}$
Social behavior	1	4	7	10	$W_{socialbehavior} = \mathbf{0{,}1}$

Parameters	$D = 1$	$D = 2$	$D = 3$
Driving skills	9	6	8
Safety	8	9	7
Primary roads usage	10	8	6
Itinerary cost	7	7	9
Social behavior	6	10	10

Fig. 3.19 Scenario 2—(**a**) parameters and respective weights, (**b**) parameter values collected through the evaluation procedure, for the 3 drivers

contrast to $\Pr\left[V_{cost} = 1 \mid D = 3\right]$ and $\Pr\left[V_{cost} = 4 \mid D = 3\right]$, which suffer a degradation, justified by the fact that the values collected from the evaluations is far higher. Again, the predominant probability $\Pr\left[V_{cost} = 10 \mid D = 3\right]$ is reinforced from the beginning and is not significantly affected after a certain number of computations.

In the following, Fig. 3.21a shows the probability density function values for the three drivers ($f\left(\bar{x}, i\right), i \in CD$). For all cases we have a level of knowledge that becomes equally high, but not at the same time: Knowledge is more quickly obtained for the last driver, as the convergence needs four iterations. In general, again, the functionality exhibits a high performance, since only a small computational effort is required for acquiring knowledge.

Decision-Making Phase

Proceeding with the second phase in this scenario, the system has to make the best driver choice, based on the passengers' preferences as depicted in the parameters' values. Figure 3.21b shows the OF values for the three drivers. As shown, the value for the third driver only needs a few iterations to become significantly higher than the others. This reveals i-CAP's ability to efficiently consider policies (preferences

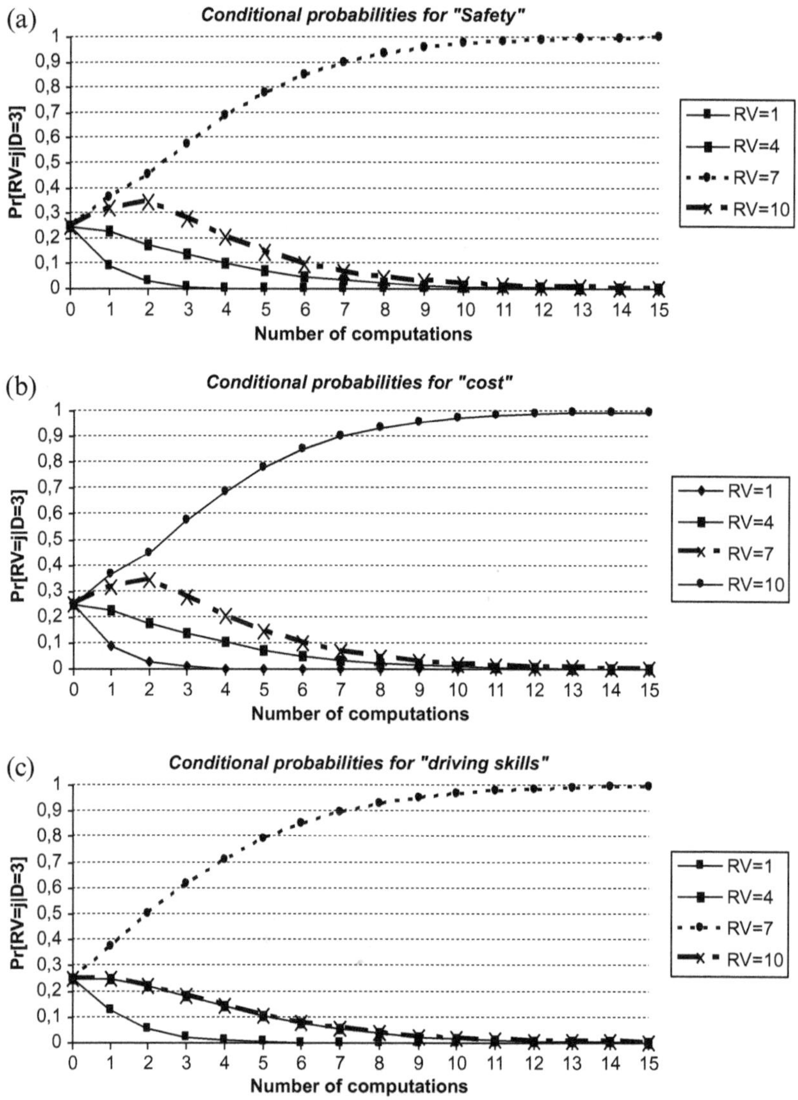

Fig. 3.20 Scenario 2, third driver—(**a**) conditional probabilities for parameter "safety," (**b**) conditional probabilities for parameter "cost," (**c**) conditional probabilities for parameter "driving skills"

depicted on weights), since the importance attributed to the overall itinerary's cost is very high and the third driver seems to be the most economically efficient one. Thus, the third driver is selected.

Scenario 3: Improvement of Driver

Scenario 3 focuses on the investigation of the system's capability to adapt to a situation that changes, meaning that at least one of the parameters' value changes significantly.

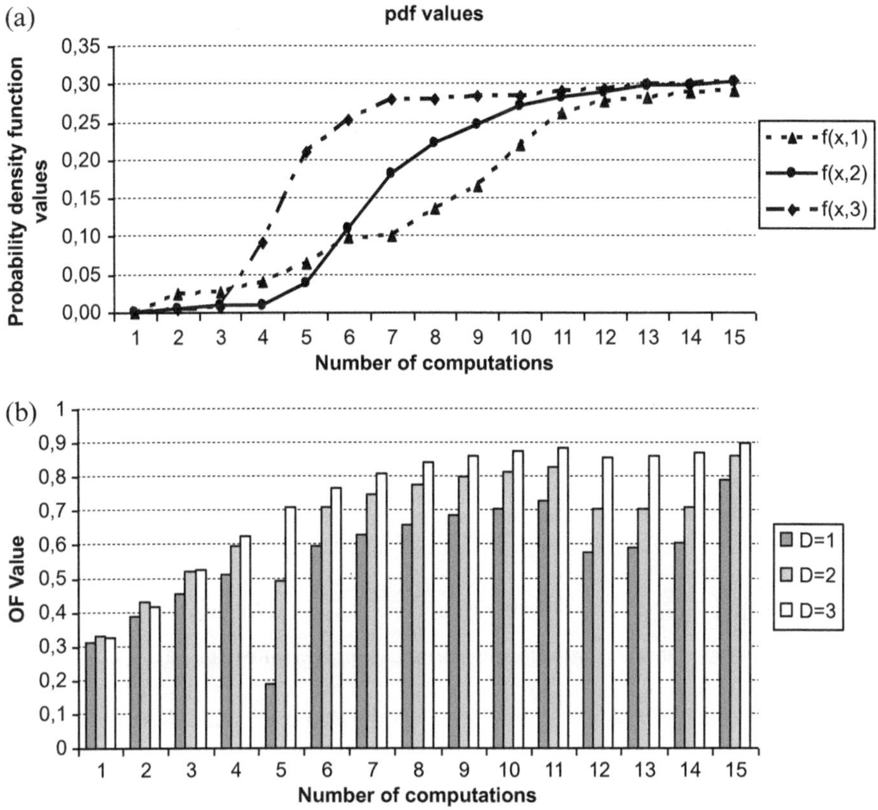

Fig. 3.21 Scenario 2—(**a**) probability density function values for the 3 drivers ($f\left(\bar{x}, i\right)$), (**b**) OF values of the 3 drivers

More specifically, in this scenario one of the drivers (the 2nd) improves his driving behavior. Figure 3.22 depicts that gradual improvement, through the evaluation marks. The weights of the parameters are the same as the ones used for the 1st scenario and are depicted in Fig. 3.16a.

In order to facilitate the process, we assume that 15 computations are divided in 3 phases consisting of 5 computations each. The second driver exhibits a better performance in each subsequent phase, when it comes to his driving skills.

Robust Discovery Phase

Figure 3.23a depicts the conditional probabilities of parameter “driving skills,” likely to be achieved by the second driver, split in 3 phases. In the first phase which lasts for 5 computations, the conditional probability $\Pr\left[V_{drivingskills} = 7 \mid D = 2\right]$ appears to be the prevalent one. Then, in the second phase (which lasts for computations 6–10), again $\Pr\left[V_{drivingskills} = 7 \mid D = 2\right]$ is the highest one. However, a slight

Parameters	$D = 1$			$D = 2$			$D = 3$		
	1st	2nd	3rd	1st	2nd	3rd	1st	2nd	3rd
Driving skills	7	7	7	6	8	9	7	7	7
Safety	8	8	8	9	9	9	6	6	6
Primary roads usage	9	9	9	8	8	8	10	10	10
Itinerary cost	6	6	6	7	7	7	8	8	8
Social behavior	10	10	10	10	10	10	9	9	9

Fig. 3.22 Scenario 3—parameter values collected through the evaluation procedure for the 3 drivers, split in three phases, namely 1st, 2nd, and 3rd

increase in the values of $\Pr\left[V_{drivingskills} = 10 \mid D = 2\right]$ is observed, with a parallel diminishment of the rest probabilities. Finally, in the third phase (computations 11–15), the most likely reference value to be achieved is 10 and thus $\Pr\left[V_{drivingskills} = 10 \mid D = 2\right]$ gradually becomes the dominant one. There is naturally a point (computation 11) where a false decision may be taken. However, the functionality quickly "recovers" and thus the small amount of time consumed for knowledge development is a desirable property. It is the time required in order to increase the reliability levels regarding the new capabilities of the second driver, in terms of achieving a satisfactory level with regard to the driving skills.

In this time period the driver exhibits a "good" behavior. In case the behavior is unstable, the improvement will be considered temporary. The different conditional probabilities will be at low levels, so they will not indicate a clear advantage for any driver. In any case, however, the amount of time required for the development of knowledge is not large, therefore enabling fast adaptations.

Decision-Making Phase

The OF values of the 2nd driver reach the highest possible values after around 12 steps on average. When possible, the 2nd driver becomes more appropriate in six steps on average. In general, a small computational effort is required for acquiring the knowledge. This is highly desirable, as it can catch improvements in the behavior of a driver even at a non-permanent basis. The number of steps is not large, and therefore, fast adaptations are possible. It should be also stated that the functionality is tested in difficult situations, since the 2nd driver becomes better in only one attribute (driving skills parameter). The time needed would be smaller if there was superiority in more attributes.

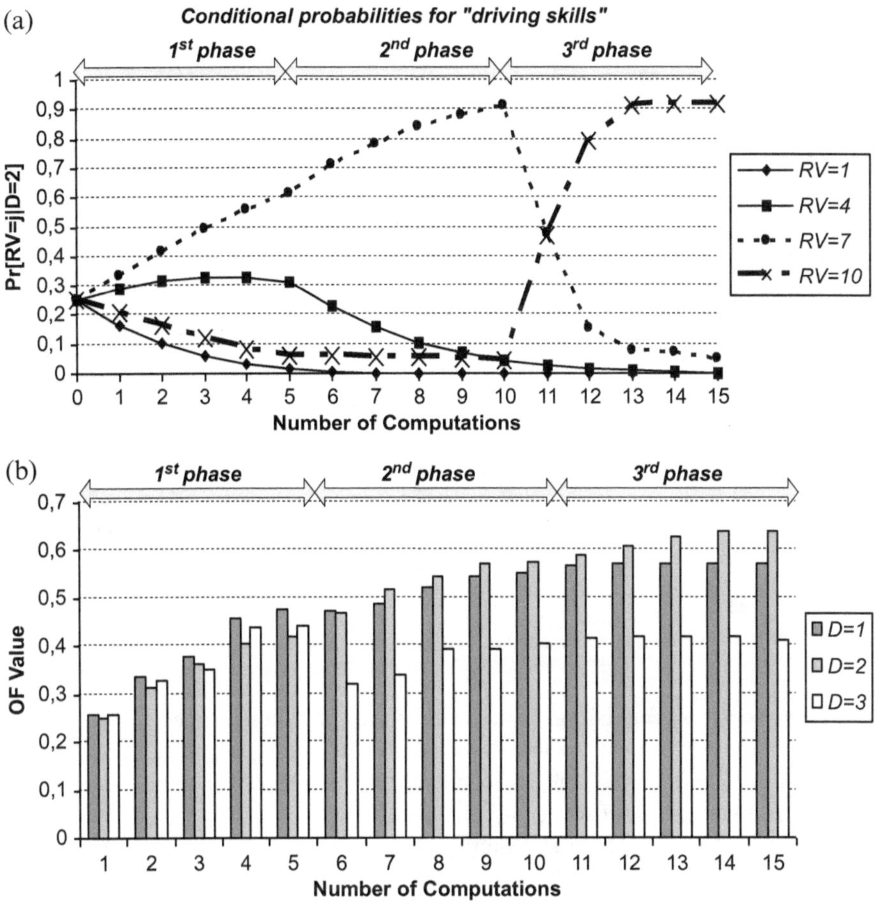

Fig. 3.23 Scenario 3—(**a**) conditional probabilities of parameter "driving skills" of the second driver in the 3 phases (the driving skills of the 2nd driver are assumed to improve in the 2nd and 3rd phases), (**b**) OF values of the 3 drivers in the 3 phases

Overall, car pooling is a representative case of an SCO that more and more cities tend to offer, often combined with most modern ideas/services, such as car sharing, as well as other urban mobility applications (see also Sect 3.1).

3.5.4 Intelligent Parking Management

A major contribution towards the improvement of the quality of transportation in large cities would be the introduction of a system that would reside inside vehicles or even consist in a smartphone application, communicate with the transportation infrastructure using IP, obtain information on "white parking spaces," and then issue the appropriate directives to the driver so as to drive the vehicle to the desired parking space.

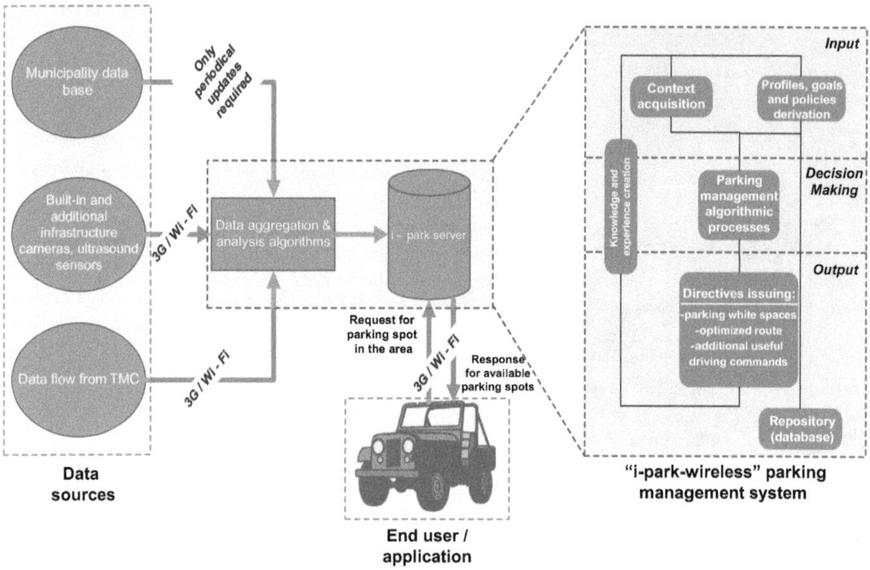

Fig. 3.24 Intelligent parking management

The information acquisition is based on a parking management system on the infrastructure side, which disposes a database of white parking spaces through information received from sensors (cameras) that constantly updates the database on the location and size of white parking spaces.

Cities tend to study more and more systems like the one shown in Fig. 3.24, since, in doing so, several impressive achievements could arise, namely:

- Minimization of the time consumed in searching for a white parking space.
- Improvement of the quality of the driver's life through a most productive utilization of time.
- Increase in the mobility efficiency through the identification of the most appropriate route towards the white parking space, adoption of "green transportation" techniques through the real-time adaptation to changes (e.g., sudden occupation of a parking space), and minimization of the GHG (Greenhouse Gas) levels in terms of the energy required/consumed from the moving vehicles.

3.6 Conclusions

This chapter has focused on the description of a high-level necessity for the development of vehicular communications and the related applications: that of sustainable mobility inside cities. As such, it has gone through the latest trends in city design, namely the one associated with the concept of "smart cities" and has presented some fundamental ideas for providing sustainable mobility solutions there in.

In conclusion, as long as economic growth is indispensably correlated with urbanization, the quest for smart and sustainable transport solutions within cities will continue to evolve. As such, it is of utmost importance for an engineer to be capable of understanding the problems that cities solved in the field of transport, and provide solutions that will make the quality of living and of moving inside cities higher.

3.7 Review Questions

Question 3.1:
What is a smart city? What is a smarter city? Can you think of examples of smart/smarter cities of today?

Question 3.2:
What are Smart City Operations and what are their main application areas ("pillars")?

Question 3.3:
Can you think of additional/future applications of Smart City Operations?

Question 3.4:
Why transport provides a fertile ground for Smart City Operations? Is this dependent on the size of a city?

Question 3.5:
What is mobility-as-a-service and "as what" has mobility been implemented so far?

Question 3.6:
Why mobility-as-a-service requires open data for its implementation?
Can you give examples?

Question 3.7:
What is car pooling and what is its difference with car sharing?

Question 3.8:
How Bayesian networks can provide the means to get an objective function to be optimized?

Question 3.9:
What are the research challenges in Smart City Operations? What are their Key Performance Indicators (KPIs) that characterize their impact?

Question 3.10:
Describe an example of a service/application that concurrently utilizes traffic management, car pooling and parking management? Is this mobility-as-a-service oriented or not necessarily?

Chapter 4
Advanced Driver Assistance Systems (ADAS)

4.1 Goals

- To present and describe to the reader the concepts of cooperative mobility and connectivity in road transport.
- To familiarize the reader with the concept of Advanced Driver Assistance Systems, as well as how they can be further advanced.
- To describe today's and tomorrows' concrete use cases/applications that fall in the realm of ADAS.
- To enable the reader to gain a holistic knowledge on ADAS, so as to be capable of designing new ADAS concepts.

4.2 Introduction

The Advanced Driver Assistance Systems are systems designed to automate, adapt, and enhance vehicle systems in order to increase safety and better driving. Safety features are based on technologies that alert the driver to potential problems, in order to avoid collisions and accidents, and at the same time they may implement safeguards and taking over control of the vehicle in order to avoid collisions. Through the implementation of adaptive features these systems may automate lighting, provide adaptive cruise control, automate braking, incorporate GPS/traffic warnings, connect to smartphones, alert driver to other cars or dangers, keep the driver in the

G. Dimitrakopoulos, *Current Technologies in Vehicular Communications*,
DOI 10.1007/978-3-319-47244-7_4

correct lane, or show what is in blind spots.[71,72,73] In general, ADAS are one of the fastest-growing segments in automotive electronics[74] since they can be based upon vision/camera systems, sensor technology, car data networks, V2V or V2I systems, whereas next-generation ADAS will increasingly leverage wireless network connectivity to offer improved value by using V2V and V2I data.

Moreover, the past decade has seen a rapidly growing interest to extend the automation of vehicles and give them the possibility to navigate the road and make tactical decisions independent of the human driver. Such *autonomous* vehicles have had a rich research history in academia, with semiautonomous demonstrations going back to the 1980s at CMU's NavLab, and notable events such as the DARPA Grand Challenges in 2004, 2005, and 2007. Since then, the technology has matured, as indicated not only by the deployment of Google's self-driving car but also by commercial plans of car vendors such as Ford, Mercedes Benz, Toyota, and Volvo, to name but a few. It is frequently argued that once adopted, these technologies will provide more efficient, comfortable, and virtually accident-free road traffic.[75,76] It is important to note that all these research programs focused on the feasibility and robustness of automated driving. Aspects related to cost are only now becoming important as commercial car manufacturers are rolling out partially autonomous vehicles. The integration with society, energy consumption, and security has received little consideration.

4.3 Cooperative Mobility and Cooperative Driving

Both economic and social vitality are highly related to mobility. Increased mobility has historically been connected to economic development, while transportation and investment in its associated infrastructure have become leading indicators of a nation's prosperity. During the last decade though, several difficulties have been

[71] G. Dimitrakopoulos et al, "A Proactive, Knowledge-Based Intelligent Transportation System based on Vehicular Sensor Networks", IET Intelligent Transport Systems journal, vol. 7, Issue:4, pp 454–463, December 2013

[72] 19. H. Fritz et al, "Chauffeur assistant: a driver assistance system for commercial vehicles based on fusion of advanced acc and lane keeping," in Intelligent Vehicles Symposium, 2004 IEEE, pp. 495–500, June 2004.

[73] J. Ploeg et al, "Design and experimental evaluation of cooperative adaptive cruise control," in ITSC, 2011 14th International IEEE Conference on, pp. 260–265, Oct 2011

[74] Ian Riches (2014-10-24). "Strategy Analytics: Automotive Ethernet: Market Growth Outlook | Keynote Speech 2014 IEEE SA: Ethernet & IP @ Automotive Technology Day". IEEE, accessed March 2015

[75] 25. M. Campbell et al, "Autonomous driving in urban environments: approaches, lessons and challenges," Philosophical Transactions of the Royal Society A, vol. 368, no. 1928, pp. 4649–4672, 2010.

[76] 26. S. Shladover, "PATH at 20-history and major milestones", "IEEE Tran. on ITS", vol. 8, no. 4, pp. 584–592, 2007

identified regarding the maintenance and improvement of mobility. Congestion exacerbated by an aging, poorly maintained infrastructure plagues many metropolitan areas. Moreover, when mobility increases usually leads to significant social concerns: the allocation of scarce land resources for transportation use, the depletion of finite energy resources, and detrimental environmental and safety impacts. In response, governments are imposing limitations on new roadway construction and the automobile itself.

Having that in mind, and based on the fact that the number of vehicles equipped with communication technology such as V2I systems (roadside units, UMTS, WiMAX, GSM) or V2V systems (IEEE 802.11p and other protocols) increase rapidly, we move towards a world where every vehicle becomes part of a collaborative ecosystem. This ecosystem can form a significant subspace of the future Internet if the heterogeneities of the underlying technologies are no longer a boundary. Moreover, the previously separated domains can cross-fertilize each other in a very beneficial way. This becomes obvious once perceived that a modern vehicle is nothing less than a mobile multi-sensor platform with over 100 different sensors. Not only can these sensors be used to monitor the driver, the vehicle, and the environment to support a safe, efficient, economic, and ecologic road transport. They form the base for a moving intelligence, detecting information and automatically inferring conclusions which eventually allow the vehicle to act on the driver's intention. Interconnecting these vehicular multi-sensor platforms with other vehicles or elements in the infrastructure using communications technology such as V2X not only does dramatically extend the "information horizon," but also allows for a whole new set of services by integrating digital and virtual entities. Cooperative applications for vehicular systems profit from additional information which cannot be gathered by in-vehicle sensor systems only. Some projects, for instance, have already explored the exploitation of position information from vehicles which are not in line-of-sight or icy road segments perceived by wheel sensors of preceding vehicles. Similarly, services known in the Internet of today will be enriched with the knowledge gained from complex situations on the road, either live or in a refined post processed quality. Learning algorithms will support the extraction of relevant behavioral or situational pattern to improve the overall quality of the transportation system on the one hand side and of the services provided in the Future Internet on the other side.

4.4 Green (eco) Driving

Novel eco-navigation applications are under development in the European R&D project eCoMove.[77] The project's approach is based on the so-called eco-cooperative horizon. The eco-cooperative horizon enables a real-time update of the

[77] eCoMove Integrated Project web site, http://www.ecomove-project.eu/, accessed December 2015

firstly calculated eco routing (= estimated less fuel consumption route), thanks to the information derived from the V2V and V2 infrastructure communication.

Specifically, such research initiatives attempt to lead to a thorough, transport-energy efficient solution, based on the development of tools that will enable drivers minimize unnecessary fuel consumption, as well as to assist road operators carry out the traffic management as energy-efficiently as possible. The implementation of such cooperative systems using vehicle-to-infrastructure and vehicle-to-vehicle communication aims to reduce fuel consumption and CO_2-emissions by 20 % overall. As such, it addresses very valuable objectives under a broad, integrative approach, following the main idea that for a given trip in a particular vehicle there is a theoretical minimum energy consumption that could be achieved by the "perfect eco-driver" traveling through the "perfectly eco-managed" road network.

As in future ITS systems the electric vehicle (EV) will become a dominant type of vehicle, research is currently focused on EVs. Some of the results will also be applicable to more traditional types of engine, such as diesel or gas fuel, or hybrid vehicles. This opens a range of applications where the EV may act as part of a smart electricity grid, or smart grid in short, as they are researched on in several projects and initiatives (such as http://portalsmartcity.sadiel.es/EN/index.html).

4.5 Connectivity in Road Transport

Connectivity in road transport is still in its infancy. As an example, a summary of the Urban Grand Challenge mentioned that a number of incidents could have been avoided if vehicles could anticipate the behavior of other vehicles, and that vehicles should cooperate in order for autonomous driving to reach its full potential.[78] The benefit of cooperation was already recognized in a parallel track in vehicle automation, namely *platooning*, which instead of complete autonomy promotes information sharing between vehicles and joint decision-making. Dating back to the 1990s and the California PATH program, the field has seen several large projects, with participants from both industry and academia.[79] In the California PATH program, congregation of vehicles in platoons with small inter-vehicle spacing was investigated as a remedy for the state's congested highways. Several European initiatives, such as the CHAUFFEUR (1999), Connect & Drive (2011), or the SARTRE (2012) projects, have similarly investigated the use of vehicle platoons, and added the objective of reduced fuel consumption. The key enabler in all projects was the reliance on vehicle-to-vehicle (V2V) and vehicle-to-infrastructure (V2I) communication for sharing information regarding the environment and internal states and choosing appropriate control policies.

---

[78] http://archive.darpa.mil/grandchallenge/, accessed March 2015.

[79] S. Shladover, "PATH at 20-history and major milestones", "IEEE Tran. on ITS", vol. 8, no. 4, pp. 584–592, 2007.

4.6 Information Sharing for Sustainable Multimodal Transport

The status of multimodal transportation networks can be acquired through several sources, some cooperative (e.g., ADS-B for flights, AIS for vessels, fleet management systems for buses and taxis, airlines, airports, railways, reservation systems, and others) and some not cooperative (e.g., traffic sensors). Currently used wireless communications networks do not allow for high refreshing rates, therefore delays of several minutes are normal ("near real time"). European flight positions can be obtained from many available sources, for instance from data feed compiled by AirNav Systems (http://www.airnavsystems.com/) from a network of interconnected positional receivers that gather ADS-B global positioning system (GPS) data transmitted from planes operating in European air space. There are similar sources for vessel positions (for instance, http://www.marinetraffic.com).

In some countries national railway and metro infrastructure managers are able to deliver near real time information about the trains. For railways, the number of traveling trains per day can reach several thousands, therefore it is not practical (nor technically feasible) today to trace the current positions, rather messages are issued to warn about delays and cancelations, broken routes, etc. Because the project objective is to deal with multimodal transportation networks, particularly in urban environments, the focus is mainly on urban railways, metros, buses, taxis, etc. while the connecting networks (flights, long range trains, ships) will be considered to allow customers to make their journeys in a smoother and more predictable manner, for instance to be timely warned about airport or railway delay information to the affected flight or train, thus avoiding to rush when not needed.

Traffic sensors networks are one of the main application areas for Wireless Sensor Networks (WSN), again a very complex area in terms of technology and standards. Sensor package (such as traffic cameras and loop detectors) normally counts passing vehicles, measures the average roadway speed, and can also be able to detect ice and water on the road

Moreover, recent development and adoption of standards for publishing information about public transportation services (e.g., GTFS, GTFS-RT, SIRI-SM, SIRI-VM) enable public transportation provider to publish information about the service routes, trips, schedule, and current locations and enable new forms of cross-model journey planning services.[80,81]

Some consideration can also be given to weather forecast agencies, and the related information networks, because adverse weather conditions (e.g., snow) are likely to heavily affect the urban traffic.

[80] A. Efentakis et al, Efficient data management in support of shortest-path computation. In Proc. 4th ACM SIGSPATIAL Workshop, CTS'11, pages 28–33. ACM, 2011

[81] Fernandez, A., Ossowski, S.A "Multiagent Approach to the Dynamic Enactment of Semantic Transportation Services", Intelligent Transportation Systems, IEEE Transactions on, vol. 12, no. 2, 2011, pp. 333–342.

About the use of social networks, there are examples where social media sites such as Facebook and LinkedIn have been leveraged to connect potential car-pool users, to deliver real-time text alerts about traffic congestion, and even to organize flexible office spaces at the entry to major highways in order to offer professionals a place to work until rush hour ends.

Moreover, at the moment, transport users have no mean to communicate to other users in the same transport system since they do not know each other.[82] This collective information could however help every user in making decisions about its transport. This is especially true in case of disturbance on the transport system (human accident, system failures, etc.). The transport company provides only delayed and incomplete information as its employees have to come on-site, analyze the situation, act, and then communicate. There are usually users which are on-site (e.g., in the train which fails, in the station where the accident happened) which know what is happening and can inform the other users on the gravity and speed of recovery.

4.7 Case Studies

4.7.1 Proactive Global Alerting Systems

As mentioned also in the previous, by enabling vehicles to communicate with each other (V2V), as well as with roadside base stations via V2I communication, intelligent transport systems can contribute to safer and more efficient roads and cities can offer smarter and smarter mobility services to their citizens and visitors.

In the light of the above, this case study is about a mobility SCO, which focuses on the management of vehicles and transportation infrastructure in a quick and efficient manner.

The proposed approach is a smart city operation indeed, as it combines (1) wireless sensors placed on the vehicles and on specific parts of the transportation infrastructure (traffic lights, road signs), (2) Wireless Sensor Networks (WSNs) formed by neighboring vehicles and parts of the infrastructure, thus referred to as "vehicular sensor networks" (VSNs), and (3) a computationally efficient heuristic for evaluating the available information and proactively issuing directives to the drivers and the overall transportation infrastructure, which may be valuable in context handling.

4.7.1.1 Formulation

In this approach we will consider a group of vehicles forming an ad hoc vehicular network with a subject vehicle and exchanging motion information among them, utilizing their embedded sensors. The subject vehicle is equipped with road

[82] V. Boschian et al, "A Metamodeling Approach to the Management of Intermodal Transportation Networks", IEEE Transactions on Automation Science and Engineering, vol. 8; p. 457–469, 2011.

network GIS data, freely available (e.g., OSM, TIGER 2014) or proprietary, and an on-board processor capable of extracting the road network information pertaining to the location area. Key to the success of this is the fast update, propagation, and exchange of the sensor information (vehicles locations, speeds, etc.) in the ad hoc network.

It is safe to assume that the peer information updates will arrive with various (short) delays to the subject vehicle. For our method to work we require a mechanism to bring the delayed positions "forward in time" and to project the peer vehicle locations accordingly. There are many methods with which we can accomplish this. From the naïve approach where the peer's speed is assumed constant and its trajectory a straight line, to a more sophisticated estimation of the speed based on previous behavior. A functional diagram of this process is shown in Fig. 4.25.

Let's accept that every vehicle is represented by a point on a plane, and all of them are dimensionless.

Let v_0 be the subject vehicle, V be the set of peer vehicles forming the ad hoc network with it at a given time, and N be their number. Therefore:

$$V = \{v_1, v_2, \ldots, v_N\}, N \in \{1, 2, \ldots\}$$

Let N be a finite set of integers such as: $N = \{1, 2, \ldots N\}$

Let v_0 be the unit vector representing the direction of the subject vehicle v_0.

The set of vectors defined as having starting point the location of the subject vehicle v_0 and ending point the location of every other vehicle $v_i, i \in N$ in the network is:

$$\overline{V_0} = \left\{\overline{v_0 v_1}, \overline{v_0 v_2}, \ldots, \overline{v_0 v_N}\right\}$$

Obviously, if there are no peer vehicles in the ad hoc network:

$$\overline{V_0} = \varnothing$$

The set of distances D between the subject vehicle and every other vehicle in the network is given by:

$$D = \left|\overline{V_0}\right| = \left\{\left|\overline{v_0 v_1}\right|, \left|\overline{v_0 v_2}\right|, \ldots, \left|\overline{v_0 v_N}\right|\right\}$$

In order to find the set of angles Θ between the subject vehicle's direction and every other vehicle in the network, we consider the following set of dot products:

$$\hat{v_0} \cdot \overline{V_0} = \left|\hat{v_0}\right| \left|\overline{V_0}\right| \cos\left(\hat{v_0} \cdot \overline{V_0}\right) = D \cos\Theta$$

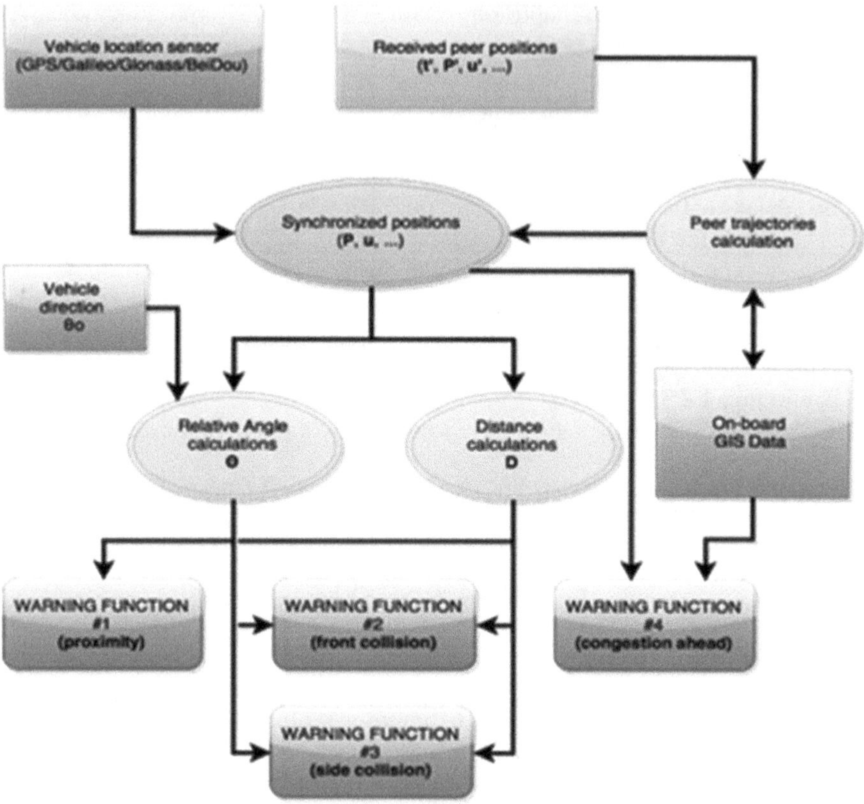

Fig. 4.25 Functional block-diagram

where every operator and function are applied to each element of the set respectively. Therefore:

$$\Theta = \arccos\left(\frac{\hat{v_0}\cdot\overline{V_0}}{D}\right) = \left\{\arccos\left(\frac{\hat{v_0}\cdot\overline{v_0\cdot v_i}}{\left|\overline{v_0\cdot v_i}\right|}\right),\ \forall i\in N\right\}$$

Before calculating D and Θ, we have to estimate the vectors $\overline{V_0}$. In order to do so, it is required to receive the geodetic coordinates of each vehicle via accurate location sensors, such as multi-constellation GPS/GLONASS/Galileo/BeiDou GNSS receivers[83] and LTE(4G), on a reference (established) datum (ellipsoid with an offset and a rotation), e.g., WGS84. For any given pair of

[83] X. Li, X. Zhang, X. Ren, M. Fritsche, J. Wickert and H. Schuh, "Precise positioning with current multi-constellation Global Navigation Satellite Systems: GPS, GLONASS, Galileo and BeiDou," Scientific Reports, vol. 5, no. 8328, Feb 2015

geodetic coordinates there is exactly one pair of Cartesian coordinates, using the established map projection, e.g., UTM.[84]

Let P be the set containing the Cartesian coordinates of every peer vehicle, such that:

$$P = \{(x_i, y_i),\ \forall i \in N\}$$

Let θ_0 be the direction (azimuth) of the subject vehicle on the Cartesian plane. The unit vector v_0 can be expressed as:

$$\hat{v}_0 = \cos(\theta_0)\hat{x} + \sin(\theta_0)\hat{y} = \begin{bmatrix} \cos\theta_0 \\ \sin\theta_0 \end{bmatrix}$$

The i-th element of the sets D and Θ can be written:

$$\mathrm{d_i} = \sqrt{(x_i - x_0)^2 + (y_i - y_0)^2},\ \forall i \in N$$

and:

$$\theta_i = \arccos\left(\frac{\begin{bmatrix} \cos\theta_0 \\ \sin\theta_0 \end{bmatrix} \cdot \begin{bmatrix} x_i - x_0 \\ y_i - y_0 \end{bmatrix}}{\sqrt{(x_i - x_0)^2 + (y_i - y_0)^2}} \right),\ \forall i \in N$$

The above expression, although mathematically accurate, may not prove to be accurate if we were to program it on a computer. There are angles near the flat extrema of the inverse cosine function where we can lose up to half the significant digits, not to mention the existence of the square root for the calculation of the distances.[85]

A more computationally accurate method for calculating the set of angles Θ would be to combine the vectors v_0 and $\overline{V_0}$ as follows:

$$\hat{v}_0 \times \overline{V_0} = \left|\hat{v}_0\right| \left|\overline{V_0}\right| \sin\left(\hat{v}_0 \cdot \overline{V_0}\right) \hat{k} = D \sin\Theta \cdot \hat{k}$$

[84] B. Jenny, "Adaptive Composite Map Projections", IEEE Trans. on Vis. and Comp. Graphics, vol. 18, iss.12, 2012, pps. 2575–2582

[85] [23] W. H. Press et al, S. A. Teukolsky, W. T. Vetterling and B. P. Flannery, Numerical Recipes, 3rd ed., Cambridge Press, 2007, pp. 1122–1124

where $\hat{k}$ the unit vector perpendicular to the 2D surface defined by vectors $\hat{v}_0$ and $\overline{V_0}$. Dividing with the dot product and substituting with the Cartesian coordinates of vectors, we get:

$$\tan\Theta = \frac{\hat{v}_0 \times \overline{V_0}}{\hat{v}_0 \cdot \overline{V_0}} = \frac{\begin{vmatrix} \cos\theta_0 & \sin\theta_0 \\ (x_i - x_0) & (y_i - y_0) \end{vmatrix}}{\begin{bmatrix} \cos\theta_0 \\ \sin\theta_0 \end{bmatrix} \cdot \begin{bmatrix} x_i - x_0 \\ y_i - y_0 \end{bmatrix}}$$

The set Θ can be derived by using the well-known atan2 function, present on most modern programming languages.

4.7.1.2 Solution: Warning Functions

At any given point in time, the sets D, Θ, and P (as well as other information that can be exchanged in the ad hoc network, referring to that time point), can be employed to create a function (warning function) F that takes the above as arguments and maps them to a set of warning level values W, as shown below. The set can be a crisp set or a fuzzy set.

$$F:(D,,,\Theta,,,P,,,,\ldots) \mapsto W, W = \{w_j,\ j = 1,2,\ldots\}$$

Since all the function arguments can be evaluated at any chosen time point, then the resulting warning level values are updated constantly over time. We will explore some of the aforementioned functions (or algorithms in broader terms), and we will present a statistical analysis using a simulated traffic environment.

Minimum Distance Warning

The goal is to issue a warning when an absolute minimum safety distance d_{safe} is breached from any direction, whilst the subject vehicle's speed u_0 is above a certain limit $u_{threshold}$, regardless of all other conditions.

Let set M be a genuine subset of N such that:

$$M = \{m \in N: u_0 > u_{threshold} \wedge d_m \in D \wedge d_m < d_{safe}\}$$

Therefore, the genuine subset of vehicles V_{WARN} that breach the safety distance is:

$$V_{WARN} = \{v_i \in V, \forall i \in M\}$$

Table 4.2 Warning level values (crisp set)

Warning level values w_j	*Safe distance*
"DANGER"	$0.8d_{safe}$
"WARNING"	d_{safe}
"CAUTION"	$1.5d_{safe}$
"OK"	$2.0d_{safe}$

The warning function is defined as:

$$F = \begin{cases} \text{"}WARNING\text{"}, V_{WARN} \neq \varnothing \\ \text{"}NOWARNING\text{"}, V_{WARN} = \varnothing \end{cases}$$

We can easily create more warning levels and corresponding messages, e.g., "DANGER," "WARNING," "CAUTION," and "OK," by using multiple appropriate safety distances, as displayed in the following crisp set, for example (Table 4.2).

We can also formulate the Warning Function outputs into a fuzzy set. There is an array of membership functions (MF) available in order to approach the construction of the fuzzy set, e.g., triangular, sigmoidal, z-shaped, and bell-shaped. We chose the sigmoidal MF and the difference of 2 sigmoidal MF.

Sigmoidal MF:

$$sigmf(x,,a,,c) = \frac{1}{1+e^{-a(x-c)}}$$

Sigmoidal Difference MF:

$$dsigmf(x,,,,a,,,,b,,,,c_1,,,,c_2) = \frac{1}{1+e^{-a(x-c_1)}} - \frac{1}{1+e^{-b(x-c_2)}}$$

We used the following membership functions in order to build the fuzzy set on Table 4.3 and its MF plot on Fig. 4.26.

Safe Following Distance Warning

The goal is to issue a warning when a safety following distance d_{front_safe} is breached.

Let u_0 be the speed of the subject vehicle at a given point in time and $t_{reaction}$ be the reaction time of a human driver, that is, the time required for perception of danger, response (e.g., decision to brake), and movement (e.g., actual application

Table 4.3 Warning level values (fuzzy set)

Warning level values w_j	*Membership function*
"DANGER"	$sigmf(x,,20,,1)$
"WARNING"	$dsigmf(x,,,,30,,,,30,,,,1,,,,1.5)$
"CAUTION"	$dsigmf(x,,,,30,,,,30,,,,1.5,,,,2)$
"OK"	$sigmf(x,,20,,2)$

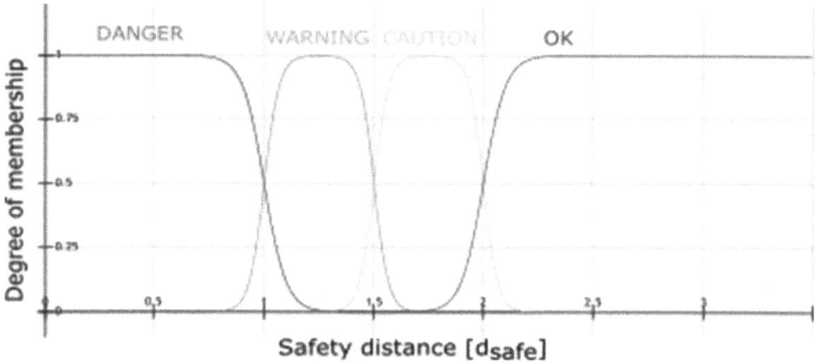

Fig. 4.26 Membership function plot diagram

of brakes). The reaction time is typically characteristic of a particular set of circumstances, including age, gender, fatigue, cognitive load, chemical usage, visual functions, illumination, etc. A maximum value of 2.5 s is often used in the United States and 2.0 s in the EU[86,87,88] with recent studies giving an average value of 1.4 s and standard deviation of 0.6 s. However, the on-board computer can make a more personalized estimate of the driver's reaction time by taking into consideration usage metrics. The reaction distance is:

$$d_{reaction} = u_0 t_{reaction}$$

If the following vehicle v_i is reducing speed (braking) to u_i, apart from the reaction distance, we need to take into account the braking distance $d_{braking}$ required by the subject vehicle in order to match its speed with the former. As such, the warnings are purely based on the speeds u_0 and u_i which are calculated for every step

[86] M. Green, "'How long does it take to stop?' Methodological Analysis of Driver Perception-Brake Times," Transp. Human Factors, 2000.

[87] X. Cui-Fang et al, "Study of driver's reaction time (DRT) during car following," Applied Mechanics and Materials, Vols. 713–715, 2015.

[88] P. M. D'Addario, "Perception-Response Time to Emergency Roadway Hazards and the Effect of Cognitive Distraction," Un. of Toronto, 2014

(position update). Therefore, even in the case when the vehicle ahead brakes, the algorithm will run and give results. We can propose that a forced position update occurs when a vehicle hits the brakes hard. We will use a baseline approach to calculate this distance, considering the rolling resistance, the air drag, and the difference in altitude as negligible throughout the length of $d_{braking}$.

Let m be the subject vehicle's mass, g be the acceleration of gravity, and μ be the coefficient of kinetic friction between the tires and the road. The required reduction in kinetic energy in order to match the following vehicle's speed is:

$$\Delta E_{kinetic} = \frac{1}{2} m \left(u_i^2 - u_0^2 \right)$$

This energy needs to be absorbed by the braking system, which in turn is required to put a work of:

$$W_{braking} = -\mu m g d_{braking}$$

which gives us the (baseline) braking distance:

$$d_{braking} = \begin{cases} \dfrac{u_0^2 - u_i^2}{2\mu g}, & u_0 > u_i \\ 0, & u_0 \le u_i \end{cases}$$

To get an idea of the order of magnitude of the kinetic friction coefficient, we refer to the two tables below (Tables 4.4 and 4.5).

The braking distance presented is of course only a baseline approach, assuming that the vehicle's brake system is able to provide an ideal braking force of $T = \mu m g$. Various inefficiencies in the braking system, such as temperature, humidity, wear, and caliper design, make the analytical calculation of the braking distance extremely difficult. For a practical implementation, pre-calculated braking distance tables can be used, which are specific to the vehicle that take into account a minimum set of the above parameters (e.g., weight, speed, road conditions), considering worst cases and applying safety margins, so that the on-board computer will be able to use a more accurate braking distance.

The minimum frontal safe distance is obtained by adding the reaction and braking distances as shown below:

$$d_{front\_safe} = d_{reaction} + d_{braking}$$

We will assume a frontal "guard" of width L, as shown in Fig. 4.27. It can be a constant value, such as a multiple of the vehicle's width, e.g., $L = 2.0 \times car\_width$

Table 4.4 Average value of tire friction coefficient

Road surface	*Peak value*	*Sliding value*
Asphalt and concrete (dry)	0.80–0.90	0.75
Asphalt (wet)	0.50–0.70	0.45–0.60
Concrete (wet)	0.80	0.70
Gravel	0.60	0.55
Earth road (dry)	0.68	0.65
Earth road (wet)	0.55	0.40–0.50
Snow (hard-packed)	0.20	0.15
Ice	0.10	0.07

Table 4.5 Tire friction coefficient on asphalt

		Road condition				
Vehicle speed (km/h)	*Tread depth (mm)*	Dry	Wet (water depth ≈ 0.2 mm)	Heavy rainfall (water depth ≈ 1 mm)	Puddles (water depth ≈ 2 mm)	Ice (black ice)
50	*New*	0.85	0.65	0.55	0.50	≤0.10
50	*1.6*	1.00	0.50	0.40	0.25	≤0.10
90	*New*	0.80	0.60	0.30	0.05	
90	*1.6*	0.95	0.20	0.10	0.05	
130	*New*	0.75	0.55	0.20	0.00	
130	*1.6*	0.90	0.20	0.10	0.00	

. It can also be an adaptive value, such as the width of the lane on which the vehicle is traveling, provided that this information is available by the on-board road network GIS data or the V2I.

Given the width L, we can calculate the maximum angular deviation θ_{front_max} from the subject vehicle's current course that would constitute a frontal obstacle, as depicted in Fig. 4.27.

As easily gathered from Fig. 4.27,

$$\theta_{front\_max} = \arctan\left(\frac{L}{2d_{front\_safe}}\right)$$

Thus, all vehicles with distances less than the safe distance and with angular deviation up to θ_{front_max} from the subject vehicle's course are considered possible forward collisions.

Similarly to (4.1) we define the set M as:

$$M = \left\{ m \in N : \begin{array}{c} d_m \in D \wedge d_m < d_{front_{safe}} \\ \wedge \\ \theta_m \in \Theta \wedge |\theta_m| \le \theta_{front\_max} \end{array} \right\}$$

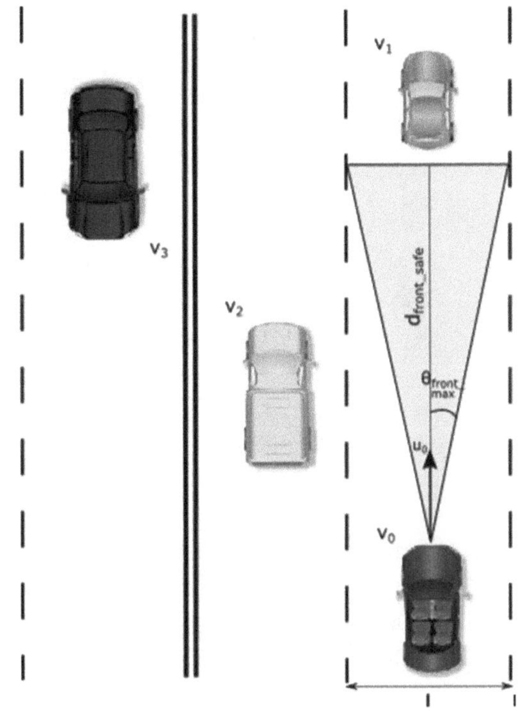

Fig. 4.27 Calculation of maximum angular deviation

We can have more warning level indications by using multiple appropriate values for d_{front_safe}, as for example in Table 4.2 (crisp set) and Table 4.3 (fuzzy set).

Safe Front-Side Distance Warning

The goal is similar to (b), with the difference in this case being that we monitor the front-side areas of the subject vehicle. We define a safe distance d_{side_safe} from each side of the vehicle (perpendicular to the vehicle's direction). This distance can be the half width of the driving lane or some other appropriate value.

With the help of Fig. 4.28 we gather that for the angular deviations θ of the peer vehicles from the subject vehicle's course are in the range:

$\theta \in (\theta_{front\_\max}, \frac{\pi}{2}]$, for the right side

and:

$\theta \in [-\frac{\pi}{2}, -\theta_{front\_\max})$, for the left side

or, altogether:

$$\frac{\pi}{2} \geq |\theta| > \theta_{front\_\max}$$

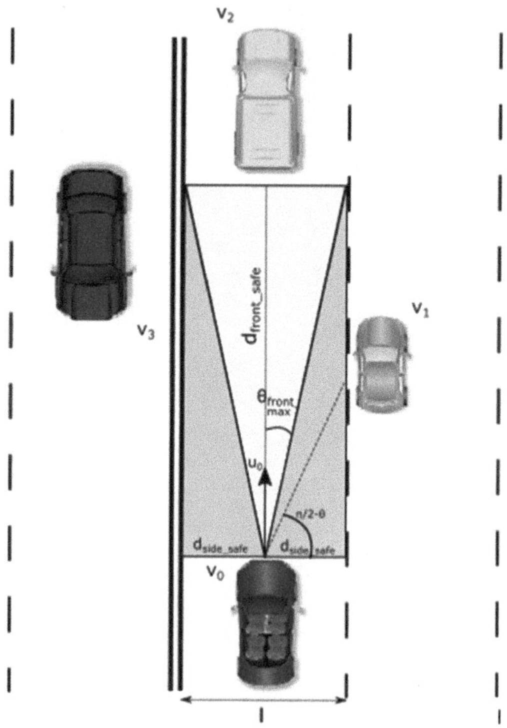

Fig. 4.28 Calculation of maximum angular deviation monitoring the front-side areas of the subject vehicle

Similarly, the minimum front-side safe distance is:

$$d_{frontside\_safe} = \frac{d_{side\_safe}}{\cos\left(\frac{\pi}{2} - |\theta|\right)}$$

Let M be a genuine subset of N such that:

$$M = \left\{ m \in N: \begin{array}{c} \theta_m \in \Theta \wedge \frac{\pi}{2} \geq |\theta_m| > \theta_{front\_max} \\ \wedge \\ d_m \in D \wedge d_m < \frac{d_{side\_safe}}{\cos\left(\frac{\pi}{2} - |\theta_m|\right)} \end{array} \right\}$$

The set of vehicles V_{WARN} that constitute safety hazard for impacts from the front-side as well as the warning function F are defined as in (4.1) and (4.2).

Congestion Ahead Warning

The goal of this warning function is to generate warning level values indicating congestion up ahead on the road. In this context, we will define congestion as the number of vehicles above a given threshold per unit of road segment area, or simply as "high vehicle density."

Consequently, the use of on-board road network GIS data, as well as any V2I information, if available, is the basis in determining the shape of the road segment ahead. Moreover, such information can also be available inside a tunnel, especially in the case of the on-board GIS computer, whereas in the case of V2I it depends on the features the tunnel offers. Specifically, we assume a road, on which the subject vehicle is traveling, of known width W and a series of sample points $S_i, i = 0,1,2,\ldots$ of the road axis, with known coordinates (x_{S_i}, y_{S_i}) . Sample point S_0 is the closest road axis sample point to the subject vehicle's location and direction at a given moment. We define the set of vectors $\overline{S}$ formed by every two consecutive sample points S_i as:

$$\overline{S} = \left\{ \overline{S_i S_{i+1}}, i = 0,1,2,\ldots \right\}$$

Let A and B be the set of points on the right and left side of the road, respectively, defined as:

$A = \{A_i\}, B = \{B_i\}$ such that $\overline{A_i B_i} \perp \overline{S_i S_{i+1}}$

Every point A_i and B_i have respective coordinates $\left(x_{A_i}, y_{A_i}\right)$ and $\left(x_{B_i}, y_{B_i}\right)$.

Figure 4.29 depicts the above concisely.

Our first goal is to find a formula to calculate the sets A and B.

Procedure:

Let:

$$\overline{s} \equiv \overline{S_i S_{i+1}} = \begin{bmatrix} x_{S_{i+1}} - x_{S_i} \\ y_{S_{i+1}} - y_{S_i} \end{bmatrix} = \begin{bmatrix} s_x \\ s_y \end{bmatrix}$$

and:

$$\overline{a} \equiv \overline{S_i A_i} = \begin{bmatrix} x_{S_i} - x_{A_i} \\ y_{S_i} - y_{A_i} \end{bmatrix} = \begin{bmatrix} a_x \\ a_y \end{bmatrix}$$

Let the length L_S (sampling length) be the road length between two consecutive samples of the axis of the road network (it is available from the on board GIS data), such that $L_S = \text{arc}\left(S_i S_{i+1}\right)$. It is not necessarily constant and usually cartographers make it dependent upon the shape of the road (e.g., linear road segments have larger sampling lengths, curvy parts have smaller sampling lengths in order to approach the shape of the road).

However, we can acquire a sufficiently small sampling length, by performing a linear or spline interpolation between the sampling points of the road segment, such that $S_iS_{i+1} \approx \text{arc}\left(S_iS_{i+1}\right)$. Therefore we have:

$$|\overline{s}| = \sqrt{s_x^2 + s_y^2} \approx L_S$$

The length of vector $\overline{a}$ is approximately equal to half the road width, given the small sampling length.

$$|\overline{a}| = \sqrt{a_x^2 + a_y^2} \approx \frac{W}{2}$$

The vectors $\overline{s}$ and $\overline{a}$ are vertical. Thus, the triangle $A_iB_iS_{i+1}$ on Fig. 4.29 is isosceles, with base length $|\overline{a}|$, height $|\overline{s}|$ and can be considered as the result of a rotation by a positive angle θ, where θ the angle i, S_iS_{i+1}, and a translation by $\left(x_{S_i}, y_{S_i}\right)$ of a triangle with its base midpoint set at $(0,0)$ and its vertex angle bisector on the positive x -axis.

The coordinates $\left(x_{A_i}, y_{A_i}\right)$ and $\left(x_{B_i}, y_{B_i}\right)$ are, respectively:

$$\begin{bmatrix} x_{A_i,B_i} \\ y_{A_i,B_i} \end{bmatrix} = \begin{bmatrix} x_{S_i} \\ y_{S_i} \end{bmatrix} + \begin{bmatrix} \cos\theta & -\sin\theta \\ \sin\theta & \cos\theta \end{bmatrix} \cdot \begin{bmatrix} 0 \\ \mp W/2 \end{bmatrix}$$

Substituting $\sin\theta \approx \frac{s_y}{L_S}$ and $\cos\theta = \frac{s_x}{L_S}$ we get:

$$x_{A_i,B_i} = x_{S_i} \pm \frac{W}{2L_S} s_y$$

and:

$$y_{A_i,B_i} = y_{S_i} \mp \frac{W}{2L_S} s_x$$

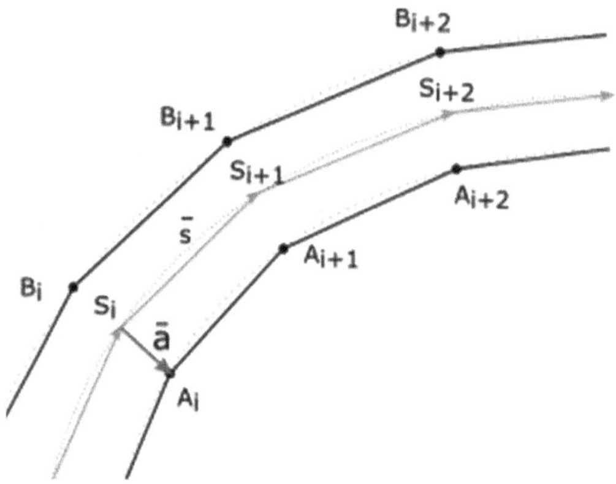

Fig. 4.29 Calculation of congestion ahead warning (1st stage)

In order to calculate the existence of traffic congestion at a road distance of L_{ahead} from the location of the subject vehicle, we will count the number of vehicles N_{window} inside a road segment "window" of length L_{window}.

Let points $S_j, S_k \in S$ be the sample points closest to the start and end of the window, respectively (Fig. 4.30).

We can derive j and k by:

$$j \in \{1,2,3,\ldots\}: \sum_{i=0}^{j-1} \left| \overline{S_i S_{i+1}} \right| \approx L_{ahead}$$

and:

$$k \in \{2,,3,,\ldots\}: j < k \wedge \sum_{i=j}^{k-1} \left| \overline{S_i S_{i+1}} \right| \approx L_{window}$$

The window (road segment) can be approximated by the simple polygon defined by the vertices:

$$Window\, polygon = A_j \ldots A_k B_k \ldots B_j$$

To make the polygon notation easier, we can define a new set of vertices C, such that:

$$C_0 \equiv A_j, \ldots, C_{k-j} \equiv A_k, C_{k-j+1} \equiv B_k, \ldots, C_{2(k-j)+1} \equiv B_j$$

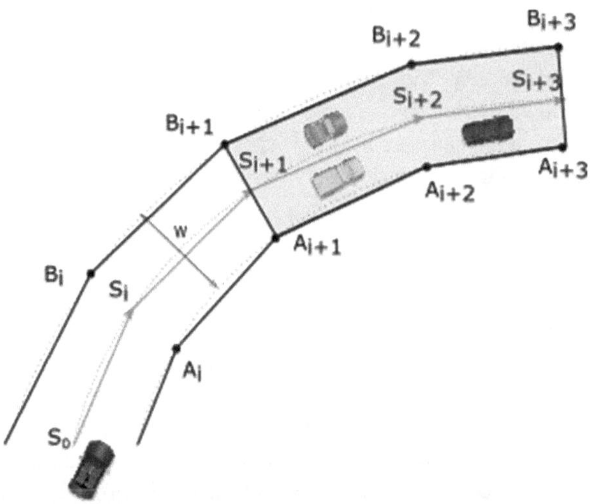

Fig. 4.30 Calculation of congestion ahead warning (2nd stage)

And so:

$$Window\,polygon = C_0 \ldots C_{2(k-j)+1}$$

The number of peer vehicles N_{Window} inside the window can be found by using an efficient "point in simple polygon" algorithm, such as the Jordan Curve Theorem.[89]

The area S_{window} of the window polygon can be calculated by the formula:

$$S_{window} = \frac{1}{2} \sum_{i=0}^{2(k-j)} \left(x_{C_{i+1}} + x_{C_i} \right) \left(y_{C_{i+1}} - y_{C_i} \right)$$

Assuming that each peer vehicle v_i exchanges its length L_{v_i} information with the subject vehicle, along with the other data, we can calculate the space occupied by every vehicle inside the window as follows:

$$S_{vehicles} = \sum_{i=1}^{N_{window}} \left(1 + c_{spacing} \right) L_{v_i} \cdot L$$

where the positive coefficient $c_{spacing}$ translates to how many car-lengths is the minimum distance ("bumper-to-bumper") between the congested vehicles in the window.

The congestion on the window of length L_{Window}, at a road distance of L_{ahead} ahead on the traveling direction of the subject vehicle, is:

$$Congestion[\%] = \frac{S_{vehicles}}{S_{window}} \times 100\%$$

The particular contribution of this SCO mainly lies in the utilization of a knowledge-based decision-making algorithm, which can increase the overall levels of safety through recognizing potential emergencies a priori, improving thus the total transportation quality. Moreover, it laterally also addresses the integration of the advantages of vehicular sensor networks in ITS through the description of a whole framework that can incorporate various services/applications that can improve the quality of transportation.

4.7.2 Reconfigurable Driving Styles

4.7.2.1 Introduction

In order to thoroughly define the driving style of a driver, we have to combine a number of characteristics, including the speed limits, the suspension adjustment, the steering wheel reciprocation, and the gear ratios. In order to change a driving

[89] W. H. Press et al, S. A. Teukolsky, W. T. Vetterling and B. P. Flannery, Numerical Recipes, 3rd ed., Cambridge Press, 2007, pp. 1122–1124

style, off-line actions are to be made: the driver, for example, may adjust one or more of these parameters with a touch of a button. On the other hand, recent technology advances in cars enable the users' online reconfiguration of their driving styles, adapting their operation to external requirements (such as the condition of the road).

Consider a case where we have a specific vehicle that may be driven by a number of potential drivers (for instance, the car of a family with several drivers), which means that we also have a number of different driving styles. The set of the drivers as well as the driving styles are related with specific parameters, like the context information that can be gathered through sensors installed in the car, data related to the personal profile of the driver, etc. Last, a set of overarching policies reflects driver/styles preferences, in the form of weights (importance) attributed to the aforementioned parameters.

The way that a driver acts and behaves when driving a vehicle may change several times. These changes are expressed with changes in the personal profile parameters. Based on that change, one of the vehicle's parameters may need to change as well. The main objective of the system is to interact with all possible driving styles (defined by the vehicle-related parameters) and choose the best for the given driver's behavior. Communication can be guaranteed through the existence of an, easy to deploy, ICT-based management system (such as the one proposed in this subsection).

The overall architecture is depicted in Fig. 4.31. As shown, the main inputs of the system comprise the personal profile parameters, the context information derived from the vehicle sensor measurements, and the policies which attribute importance to the parameters through numerical weights. The system's output is the optimum matching among drivers and driving styles. The procedure in order to get this output consists of two discrete phases: the "robust discovery phase" and the "decision-making phase." The robust discovery phase aims at maximizing the probabilities that the parameters will reach certain values, through a Bayesian-based model, which helps the system obtain knowledge. The decision-making phase steps on those probabilities and finds the optimum matching considering also the importance of the parameters.

It should be also noted that knowledge acquisition is further enhanced by an evaluation procedure, made by the driver concerning driving styles after the completion of a ride. In this respect, parameters are evaluated, at an integers' scale from "1" to "10," in the form of utility volumes, with "1" standing for "poor" and "10" standing for "excellent."

4.7.2.2 Formal Description

- The set of the potential vehicle's drivers is *PD*.
- *D* is defined for representing the driver. *D* can take values 1 to $|PD|$.
- The set of candidate driving styles is denoted as *CDS*.

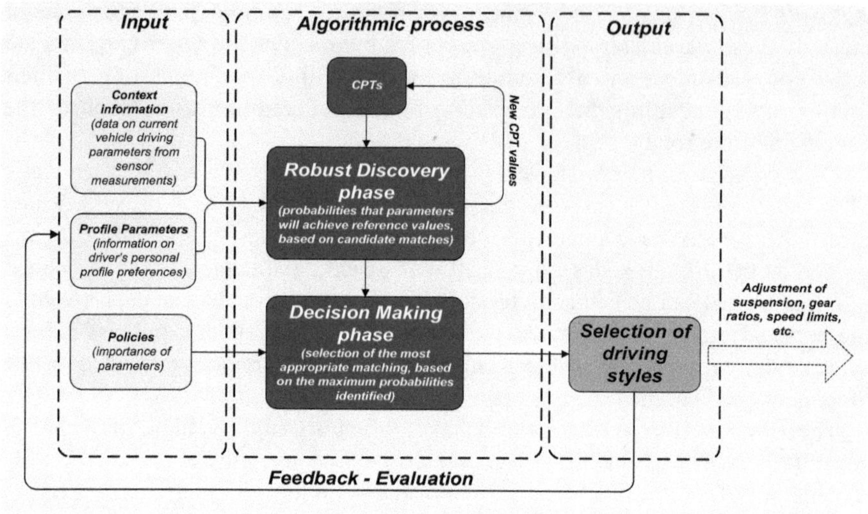

Fig. 4.31 Reconfigurable driving styles—high-level description

- *DS* is defined for representing the driving style. *DS* can take values 1 to |*CDS*|.
- The set of parameters is denoted as *N*.
- Each parameter, j $(j = 1,\ldots,N)$, can refer to a specific aspect, e.g., mean driving speed, age, and gender. Finally, the importance of each parameter, j $(j = 1,\ldots,N)$ is indicated by a weight value w_j.
- The sum of the w_j weights, over all $j = 1,\ldots,N$, will be 1.
- Variable i is defined for representing the driving style.
- Variable v_j $(j = 1,\ldots,N)$ depicts the value of the j-th parameter.
- Each variable V_j is associated with a set of reference values RV_{ij} $(i \in CDS)$.
- Variable v_j can take a value among those in RV_{ij}, when driving style i is considered.
- The knowledge that needs to be developed relies on conditional probabilities, which have the form $\Pr\left[V_j = rv_{ij}^{k} \mid DS = i\right]$, where $rv_{ij}^{k} \in RV_{ij}$ denotes the k-th reference value for the j-th parameter when driving style i is considered.

Probability density function. The following probability density function can be defined:

$$f\left(\tilde{x}, i\right) = \Pr\left[V_1 = rv_{i1}^{k1}, \ldots, V_N = rv_{iN}^{kN}, DS = i\right]$$
$$= \Pr\left[DS = i\right] \cdot \prod_{j=1}^{N} \Pr\left[V_j = rv_{ij}^{kj} \mid DS = i\right] \quad (4.1)$$

where $i \in CDS$, $\tilde{x} \in X_i$, $rv_{ij}^{kj} \in RV_{ij}$ ($j = 1,\ldots,N$), and k_j ($j = 1,\ldots,N$) are integers.

The $\Pr[DS = i]$ probabilities show the volume of information existing for each driving style *i*. The sum of the $\Pr[DS = i]$ quantities, over all $i \in CDS$, is 1. The more information there is on a driving style *i*, the more reliable the knowledge, and therefore, the higher the $f\left(\tilde{x},i\right)$ values.

In general, the values of the $f\left(\tilde{x},i\right)$ function express in an aggregate manner our knowledge on how probable is the achievement of a parameter value indicated in x, by driving style *i*.

- The goal of this process is to identify the most probable parameter values.
- For this purpose, *a-drive* collects evaluations made for the *CDS* driving styles.
- Let us assume that the most recent evaluation indicates that driving style *i* can achieve rv_{ij}^{coll} regarding parameter *j*.
- Let dif_{ij} be the difference between the maximum and the minimum reference value in RV_{ij}.
- Then, for each reference value, $rv_{ij}^{k} \in RV_{ij}$, there can be a correction factor
- $cor_{ij}^{k} = 1 - \left(\left|rv_{ij}^{k} \;\; rv_{ij}^{coll}\right| / dif_{ij}\right))$.
- Since $0 \le cor_{ij}^{k} \le 1$, a value close to 1 means that the reference and collected values are close
- New conditional probabilities

$$\Pr\left[V_j = rv_{ij}^{k} \middle| DS = i\right]_{new} = nf_{ij} \cdot cor_{ij}^{k} \cdot \Pr\left[V_j = rv_{ij}^{k} \middle| DS = i\right]_{old}$$

Parameter nf_{ij} is a normalizing factor for guaranteeing that all the "new" probabilities will sum up to one

Summing up, the updating procedure comprises collecting the parameter reference values, computing the correction factors through relation, and the new probabilities through relation, then check if a probability exceeds pr_{max} and set it equal to the threshold, and finally calculate the new normalizing factors, by forcing the remaining probabilities to sum to ($1 - pr_{max}$), and the new values are computed for the remaining probabilities.

Decision-Making Phase: Exploitation of Knowledge

This phase focuses on the choice of the optimal driving style, which will lead to the optimization of the parameter values as well. For that we need the definition of an Objective Function (OF) value, OF_i, that is calculated for every driving style, $i \in CDS$. The calculation of the OF values, OF_i, for every driving style $i \in CDS$, is carried out using the following equation:

$$OF_i = \sum_j \left\{ \max\left(\Pr\left[V_j = rv_{ij}^{k} \mid DS = i\right]\right)\right\} \cdot w_j \qquad (4.2)$$

where $i \in CDS$, ($j = 1,\ldots,N$) and $rv_{ij}^{k} \in RV_{ij}$ denotes the *k*-th reference value for the *j*-the parameter when driving style *i* is considered. The driving style with the highest OF_i value is then chosen.

4.7.3 Video-Based DAS

A critical part of an ADAS is that the system has to evaluate image sequences recorded with cameras mounted in a moving vehicle. These images can give information about the context related to the vehicle. The analysis of this information is required in order for that to be able to support and enhance driver in real traffic cases. A thorough presentation on the development of image sequence analysis systems for road vehicles can be found in Enkelmann (1997).[90]

For instance, the presence, speed, and type of vehicles can be detected based on the infrared energy radiating from the detection area. The main drawbacks are the performance during bad weather and limited lane coverage.

Video image detection is also used in traffic flow measurement: video cameras record vehicle numbers, type, and speed by means of different video techniques, e.g., trip line and tracking (Fig. 4.32).

Video image detection can also be used for functions based on vehicle identification like parking garage control, highway toll collect control, or average speed detection. Long-term average speed detection is based on the identification (number plate) of the vehicle at different fixed points of the infrastructure (e.g., highway). As the distance between the fixed measuring points are known, the average vehicle speed can be calculated from the time of appearance. The system can be sensitive to meteorological conditions.

4.7.4 Radar-Based DAS

The technology of microwave radar is able to detect moving vehicles and speed, while it is not affected by weather conditions. Fixed or portable radar (LIDAR) systems are used, for example, by the police for speed control of the vehicles; also, local communities are operating the same technology for setting up fixed speed warning signs at the entrance of the city (village) or just before a dangerous curve.

Another alternative is the technology of ultrasonic and passive acoustic. Such devices use sound waves in order to detect vehicles; they actually measure the time that the signal needs in order to reflect to the vehicle and return to the device. The ultrasonic sensors are placed over the lane and can be affected by

[90] Enkelmann, W. 1997. Entwicklung von Systemen zur Interpretation von Straßenverkehrsszenen durch Bildfolgenauswertung. Habilitation, Fakultat f ¨ ur Informatik, Universit ¨ at Karlsruhe (TH), Juli 1996. Infix-Verlag, Sankt Augustin

Fig. 4.32 Highway toll control cameras (Source: www.nol.hu)

temperature or bad weather. The passive acoustic devices are placed alongside the road and can collect vehicle counts, speed, and classification data. They can also be affected by bad weather conditions (e.g., low temperatures, snow) (Fig. 4.33).

4.7.5 Head-up Display-Based DAS

Head-up display (HUD) is a rather old technology, first implemented in aircrafts. A typical HUD contains three primary components: a projector unit, a combiner, and a video generation computer.

The projection unit in a typical HUD is an optical collimator setup: a convex lens or concave mirror with a Cathode Ray Tube, light emitting diode, or liquid crystal display at its focus. This setup (a design that has been around since the invention of the reflector sight in 1900) produces an image where the light is collimated, i.e., the focal point is perceived to be at infinity.

The combiner is typically an angled flat piece of glass (a beam splitter) located directly in front of the viewer, which redirects the projected image from projector in such a way as to see the field of view and the projected infinity image at the same time. Combiners may have special coatings that reflect the monochromatic light projected onto it from the projector unit while allowing all other wavelengths of

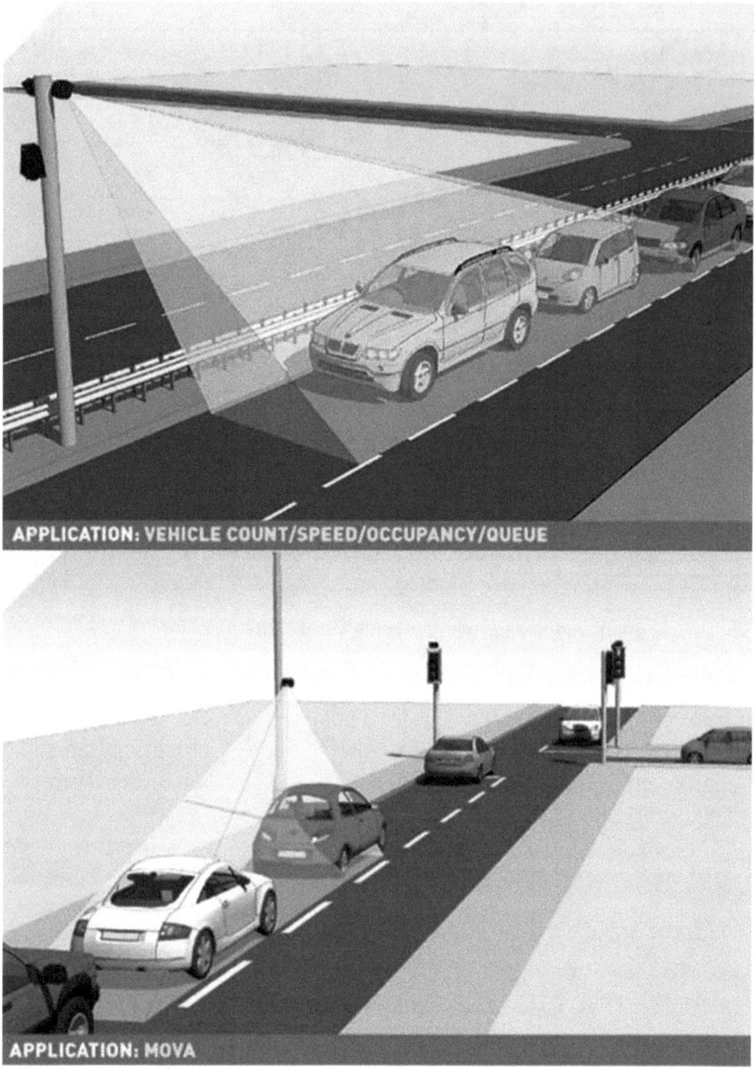

Fig. 4.33 Radar-based measurement solution (Source: http://www.roadtraffic-technology.com, AGD Systems)

light to pass through. In some optical layouts combiners may also have a curved surface to refocus the image from the projector.

HUDs are split into four generations reflecting the technology used to generate the images. In the 1st generation systems, a CRT was used in order for an image to be generated on a phosphor screen. The main disadvantage in this case was the phosphor screen coating degrading over time. Nevertheless, most of the HUDs available today belong to this generation.

The 2nd generation systems use a solid state light source, for example, LED, which is modulated by an LCD screen to display an image. These systems do not fade or require the high voltages of first generation systems. These systems are on commercial aircraft.

In the 3rd generation systems optical waveguides are used in order to produce images directly in the combiner rather than use a projection system. Finally, regarding the 4th generation systems, a scanning laser is used in order to display images and even video imagery on a clear transparent medium.

HUD-based DAS in cars have recently become very popular, first in the USA and Asia and later in Europe. These displays are becoming increasingly available in production cars, and usually offer speedometer, tachometer, and navigation system displays. Night vision information is also displayed via HUD on certain automobiles. In 2012 Pioneer Corporation introduced a navigation system that projects a HUD in place of the driver's visor that presents animations of conditions ahead, a form of augmented reality (AR).[91] Add-on HUD systems also exist, projecting the display onto a glass combiner mounted on the windshield.

Moreover, recent initiatives enable augmented reality solutions inside vehicles. Indicatively, as shown in Fig. 4.34, a Korean company has developed a DAS that allows the driver to view the actual image of the road and see a range of useful traffic information.

4.7.6 Driver Fatigue Detection Systems

Fatigue is the physical and mental impairment brought about by inadequate rest over a period of time. Ideally, people need 7/8 h sleep every night. Drivers suffering from a sleep debt are at risk of "nodding off" whilst driving and substantially increasing their risk of being involved in a crash. It is estimated that driver fatigue is a contributory factor in as many as 1 in 5 driver deaths in Ireland every year. Furthermore, tiredness-related collisions are 3 times more likely to be fatal or result in a serious injury because of the high impact speed and lack of avoiding action. A survey of drivers' attitudes to driver fatigue conducted by the RSA in 2014 revealed that over 1 in 10 motorists have fallen asleep at the wheel. The survey also found that motorists who drive as part of their work, and motorists who admit to driving after taking any amount of alcohol, had a higher than average incidence of falling asleep at the wheel (almost 1 in 5 fell asleep at the wheel).

To tackle such situations, driver fatigue detection systems have been early implemented in the context of various initiatives, with the aim to prevent accidents caused by the driver getting drowsy.

[91] Pioneer launches car navigation with augmented reality, heads-up displays System also uses dash cams to share images of street conditions across Japan. Alabaster, Jay | Computerworld | Pioneer launches car navigation with augmented reality, heads-up displays June 28, 2013

Fig. 4.34 Head-up display-based DAS. https://e27.co/korean-in-car-navigation-startup-launches-augmented-reality-driving-system-20141230/

The fatigue/inattention/drowsiness are very vague concepts. These terms refer a loss of alertness of vigilance while driving. Indicators of fatigue can be found in Bittner et al. (2000).[92]

4.7.6.1 Visual Features

There is an important quantity of studies related with this area.[93] Most of them are based on facial recognition systems to determine the position of the driver's head, the frequency of blinking, etc.

This frequency and the degree of eyelid opening are good indicators of tiredness level.[94] In a normal situation, driver blinks and moves the eyes quickly and constantly, keeping a large space between eyelids. In a sleepy state, we can appreciate that the speed of blinking and the opening decrease.

With regard to the driver's head angle in a normal situation, he maintains a lifted up position and only does the typical movements related to the driving. Passing into a drowsy state implies to nod off as well as a more frequent head's position change.

[92] R. Bittner, K. Hána, L. Poušek, P. Smrka, P. Schreib and P. Vysoký, Detecting of Fatigue States of a Car Driver, ISMDA 2000, LNCS 1933, pp. 260–274, 2000.

[93] G. Scharenbroch, Safety vehicles using adaptive interface technology (SAVE-IT) (Task 10): Technology review. Delphi electronics and safety systems Tech. Rep., 2005

[94] Qiang Ji, Zhiwei Zhu, y Peilin Lan: Real-Time Nonintrusive Monitoring and Prediction of Driver Fatigue. IEEE Transactions on Vehicular Technology, vol. 53, n°. 4, July 2004

In fact, when it is a deep stage, the nodding off is extremely slow and the head keeps itself completely relaxing.[95]

Other research lines are centered in the analysis about facial expression. In general, people are prone to have different expression depending on the alert level that show.[96]

4.7.6.2 Nonvisual Features

Driver's concentration can be affected by environmental factors; therefore it would be interesting to sensorize the cabin. Diverse studies analyze the concentration of carbon monoxide and oxygen in air. An intelligent gas sensing system offers an added security in the vehicle, warning when the concentration is higher than tolerable levels (CO of 30 ppm and oxygen levels below 19.5 %).[97]

Other nonvisual features are physiological variables. Galvanic skin response (GSR) and the conductivity are relation to the psychological state of the person.[98]

Gripping force gives us an idea about driver's attention level, and body temperature is an important physiological parameter that depends on driver's state too: body temperature increases due to infections, fever, etc., reflecting the autonomic responses and the activity of a human's autonomic nervous system.[99] Electroencephalogram gives a lot of psychophysiological information about stress state, drowsiness, or emotional reactions.[100]

4.7.7 *Obstacle Recognition*

Reliably and accurately detecting obstacles is one of the core problems that need to be solved to provide advanced safety management in vehicles. For many use cases such as micro aerial vehicles (MAVs) or self-driving cars, obstacle detection

[95] Mai Suzuki, Nozomi Yamamoto, Osami Yamamoto, Tomoaki Nakano, y Shin Yamamoto: Measurement of Driver's Consciousness by Image Processing-A Method for Presuming Driver's Drowsiness by Eye-Blinks coping with Individual Differences—2006 IEEE International Conference onSystems, Man, and Cybernetics

[96] Haisong Gu, Qiang Ji†, Zhiwei Zhu: Active Facial Tracking for Fatigue Detection. Applications of Computer Vision, 2002. (WACV 2002)

[97] Galatsis, K.; Wlodarski, W.; Li, Y.X.; Kalantar-zadeh, K.: Vehicle cabin air quality monitor using gas sensors for improved safety. Optoelectronic and Microelectronic Materials and Devices, 2000. COMMAD 2000. Proceedings Conference on Volume, Issue, 2000 Page(s): 65–68

[98] F. Nasoz, O. Ozyer, C. L. Lisetti, and N. Finkelstein, "Multimodal affective driver interfaces for future cars," in Proc. ACM Int. Multimedia Conf. Exhibition, 2002, pp. 319–322

[99] Axisa, F. Dittmar, A. Delhomme, G. : Smart clothes for the monitoring in real time and conditions of physiological, emotional and sensorial reactions of human. Microcapteurs et Microsyst. Biomedicaux, INSA Lyon, Villeurbanne, France;

[100] Lin, Y. Leng, H. Yang, G. Cai, H.: An Intelligent Noninvasive Sensor for Driver Pulse Wave Measurement. Sensors Journal, IEEE

approaches need to run in (near) real time so that evasive actions can be performed. At the same time, solutions to the obstacle detection problem are often restricted by the type of vehicle and the available resources. For example, a MAV has restricted computational capabilities and can carry only a certain payload while car manufacturers are interested in using sensors already built into series vehicles in order to keep self-driving cars affordable.

There essentially exist two approaches for obstacle detection. Active methods use sensors such as laser scanners, time-of-flight, structured light, or ultrasound to search for obstacles. In contrast, passive methods try to detect obstacles based on passive measurements of the scene, e.g., in camera images. They have the advantage that they work over a wide range of weather and lighting conditions, offer a high resolution, and that cameras are cheap. At the same time, a wide field of view can be covered using, for example, fisheye cameras.

Active sensors such as lidar[101] and binocular stereo cameras[102] can provide a depth map of the environment at any time, even when the vehicle is not moving. Stereo cameras offer the advantage of being cheap to produce while providing high-quality measurements in real time. Thus, many obstacle detection systems rely on a stereo setup.[103] Obstacles are usually detected in an occupancy grid.[104]

4.7.8 Distraction Detection

Although most motor-vehicle crashes are attributed to multiple causes, driver error represents a dominant one because drivers are responsible for operating vehicles and avoiding crashes.[105] Compared to 34.9 % for roadway factors and 9.1 % for vehicle factors, driver errors contribute to 92.9 % of crashes.[106] For example, rear-end collisions that comprise approximately 30 % of all crashes and roadway departure crashes, which cause the greatest number of fatalities have been largely attributed to the inability of drivers to detect hazards and control the vehicle properly.[107]

---

[101] Y.-W. Seo, "Generating omni-directional view of neighboring objects for ensuring safe urban driving," Carnegie Mellon University, Tech. Rep., 2014.

[102] A. Geiger, J. Ziegler, and C. Stiller, "Stereoscan: Dense 3d reconstruction in real-time," in Intelligent Vehicles Symposium (IV), 2011

[103] U. Franke, D. Pfeiffer, C. Rabe, C. Knoeppel, M. Enzweiler, F. Stein, and R. Herrtwich, "Making Bertha see," in ICCV Workshop Computer Vision for Autounomous Vehicles, 2013.

[104] A. Elfes, "Sonar-based real-world mapping and navigation," Journal of Robotics and Automation, 1987

[105] Lee, J. D. (2006). Driving safety. In R. S. Nickerson (Ed.), Review of Human Factors. Santa Monica, CA: Human Factors and Ergonomics Society.

[106] Treat, J. R., Tumbas, N. S., McDonald, S. T., Shinar, D., Hume, R. D., Mayer, R. E., et al. (1977). Tri-level study of the causes of traffic accidents (No. DOT-HS-034-3-535-77): Indiana University

[107] The National Safety Council. (1996). Accident Facts. Itasca, IL.

Most of these performance breakdowns result from the impairments of driver's attention. Four major categories of attentional impairments include alcohol, fatigue, aging, and distraction. Alcohol contributes to approximately 40 % of fatalities in US highway. Fatigue is often cited in the accidents involving young drivers and truck drivers because these drivers tend to adopt risky strategies to drive at night and/or lack good-quality sleep. Aging results in longer response time to hazards and more narrow field of attention in old drivers.

Compared with the above three impairments, distraction, the fourth impairment, is the impairment that has become increasingly important with the introduction of in-vehicle technology (e.g., navigation systems, cell phones, and internet) and has drawn increasing attention from human factor researchers and policy makers in the area of transportation safety. Driver distraction diverts driver's attention away from the activities critical for safe driving toward a competing activity.[108] It contributes to 13–50 % of all crashes, resulting in as many as 10,000 fatalities and $40 billion in damages each year. In the 100-Car Study[109] driver inattention contributed to nearly 80 % of the crashes and 65 % of the near-crashes. The trend toward increasing use of in-vehicle information systems (IVISs) is critical because IVISs induce distraction, which includes two major types: visual distraction and cognitive distraction.

Visual distraction can be described as "eye-off-road," and cognitive distraction as "mindoff-road."[110] Both types of distraction can lead to larger lane variation, more abrupt steering control, slower response to hazards, and less efficient visual perception than attentive driving. Moreover, these two types of distraction can occur in combination and interact with each other.

A promising strategy to minimize driver distraction is to develop adaptive distraction mitigation systems, which adjust their functions and provide assistance to reduce distraction based on the state of drivers. These systems build on the concept of cooperative automation, which includes identifying the user's state and adapting to it.

Such as system usually assesses driver state based on the information collected by a range of sensors in real time and provides the driver with mitigation strategies to maintain acceptable performance. The mitigation strategies in such an adaptive system include warning drivers, blocking distraction sources, and/or providing feedback. For example, the SAVE-IT (Safety VEhicle using adaptive Interface Technology) project developed a vehicle incorporating adaptive interface technology to mitigate driver distraction and evaluated its safety benefits.[111]

---

[108] Lee, J. D., Young, K. L., & Regan, M. A. (2008). Defining driver distgraction. In M. A. Regan, J. D. Lee & K. L. Young (Eds.), Driver distraction: Theory, effects, and mitigation (pp. 31–40). Boca Raton, FL: CRC Press, Taylor & Francis Group.

[109] Klauer, S. G., Neale, V. L., Dingus, T. A., Ramsey, D. J., & Sudweeks, J. (2005). Driver inattention: A contributing factor to crashes and near-crashes. Paper presented at the Human Factors and Ergonomics Society 49th Annual Meeting, Orlando, FL

[110] Victor, T. W., Harbluk, J. L., & Engström, J. A. (2005). Sensitivity of eye-movement measures to in-vehicle task difficulty. Transportation Reseach Part F, 8, 167–190

[111] Witt, G. J. (2003). SAfety VEhicle(s) using adaptive Interface Technology (SAVE-IT) Program.

4.7.9 Lane Keeping and Lane Departing

Lane departure warning systems are designed to help reduce the likelihood of the vehicle leaving the road or crossing into an oncoming lane and therefore the risk of accident as a result of driver distraction or a lapse in concentration. Such a system usually monitors the road ahead with the aid of a camera (located near the interior rear-view mirror) which recognizes lane markings and evaluates the position of the vehicle. If the vehicle starts to leave the lane, the system takes corrective steering action or vibration depending on model. If this is not sufficient, the driver is warned about the situation by a steering vibration and is asked to take over the steering. Additionally, if no active steering movements by the driver are recognized for longer than approximately 8 s, a message will appear in the Multi-Function Display in conjunction with a warning tone. The corrective steering function can be overridden by the driver at any time and the system does not react if the turn indicator is set before crossing a lane marking. Lane Assist cannot replace the driver's attentiveness. The driver is still legally responsible for the vehicle and therefore staying in the lane at all times. The system will not work if there are no recognizable lane markings. The camera vision can be reduced by rain, snow, heavy spray, or oncoming lights. This and vehicles in front of you can lead to the lane markings not being recognized by the system.

In general, such systems are not activated at a vehicle speed of less than 65–70 km/h.

4.7.10 Proactive Emergency Braking

Proactive emergency braking systems use a combination of cameras, lasers, and radar to detect another vehicle in front of your car. If such a system senses a crash is about to happen, it applies the brakes—unless you do so first. This is especially effective in rear-end collisions. The technology behind a proactive emergency braking system is among many that will eventually appear on autonomous vehicles, according to automotive observers.

Several researchers and experts claim that such systems can and do fail from time to time. They urged automakers to make sure their cars are otherwise safe and offer protection to passengers in the event of a failure. Experts also said systems of this kind are not a substitute for regulations, such as those prohibiting tailgating or insufficient assured clear distance between vehicles.

In addition to concerns expressed by safety experts, some consumer groups, such as Ralph Nader's Center for Auto Safety, want to see enforceable regulations instead of an agreement. These groups point to the General Motors ignition-switch problem and Toyota brake issues as proof that absent strict laws, automakers may try to hide defects or choose simply not to comply.

The leap to adopt emergency braking systems as a standard feature won't be large for some automakers that already offer automatic emergency braking as an option in some of their lines. In fact, back in September 2015, Audi, BMW, Ford, General Motors, Mazda, Mercedes-Benz, Tesla, Toyota, Volkswagen, and Volvo said they planned to make such a standard in their full line of cars and trucks.

4.7.11 *Remote Vehicle Monitoring*

Remote vehicle monitoring has been traditionally approached by using embedded vehicle monitoring devices that send info to remote stations, other vehicles, and infrastructure. As the need for real-time information is growing, it is important to connect all those devices to the communication networks (via satellite or GPRS) to collect all the information in control centers to allow a better vehicle management, traffic planning, and to solve problems faster and proactively, which leads to cost savings and more efficient operations management.

Several manufacturers have already announced those products, such as Volvo CareTrack, Mercedes FleetBoard, Delphi's Verizon,[112] or GuardMagic. However, most of those solutions pose four challenges: it is very slow to adapt all floats; there is no standard solution which makes integration very difficult in an open ITCS; and, even if adaptable, they are expensive for many SME because of all the special equipment and installation required. However, with the increasing demand for transportation for all modes of transportation and types of vehicles, it is very important to adapt ITCS to cope with Automatic Vehicle Location and Monitoring Systems (AVL and AVM) coming from several manufacturers without incurring in high-budget expenses.

It is then important to have a low adaptation budget and to involve public–private partnerships to make the system widely available, even in some cases at the price of reducing the information coming from vehicles and infrastructure if all facilities are not present. Examples of European projects are the Ventotene project, that aims to develop and test an Automatic Vehicle Monitoring and Location System in the Ventotene Italian island using GPS, EBSF (European Bus System of the Future), and ECOMPASS (eCO-friendly urban Multi-modal route PlAnning Services for Mobile uSers). All of them are thought for specific sectors and integration of data systems is not foreseen.

4.8 Conclusions

This chapter has focused on one of the most prominent fields that is associated with vehicular communications, namely that of ADAS. ADAS promise significant advantages and novelties to the manner in which we drive vehicles, facilitating

[112] Jenkins, W., Lewis, R., Lazarou, G. Y., Picone, J., & Rowland, Z. (2004, November). Real-time vehicle performance monitoring using wireless networking. In Communications, Internet, and Information Technology (pp. 375–380).

several of the driver's operations and the passengers' journey, as well as protecting the vehicle from undesired situations.

ADAS constitute a field that continues to attract immense research. Furthermore, it is expected that with the advent of autonomous driving solutions (as explained in Chap. 5), the research and development work in ADAS will not only be continued, but will also be probably intensified, so as to holistically undertake the responsibility of getting a vehicle safely from point A to point B.

4.9 Review Questions

Question 4.1:

What are Advanced Driver Assistance Systems? Do they assist only the driver or also passengers and/or other transport stakeholders?

Question 4.2:

Why are we still studying ADAS since they have been in the market since the 1990s?

Question 4.3:

Are ADAS related exclusively to the utilization of ICT and in-vehicle electronics or not?

Question 4.4:

What are the main dimensions (areas of application) of ADAS?

Question 4.5:

Can you think of future ADAS oriented services/applications?

Question 4.6:

Do ADAS require always in-vehicle connectivity or they are about embedded intelligence inside vehicles?

Question 4.7:

Can you describe how can angles be calculated in estimating the oncoming danger for collision inside a GPS-equipped vehicle?

Question 4.8:

Can you think of an application scenario for each of the ADAS areas presented in Sects. 4.7.1 up to 4.7.11?

Chapter 5
ICT-Enabled, Knowledge-Based (Cognitive) Management Algorithms for ADAS

5.1 Goals

- To familiarize the reader with the communication enablers for making V2I/V2I a reality.
- To present the requirements and advances for V2V/V2I management functionality.
- To describe ICT-enabled management functionality approaches for supporting the operation of ADAS.
- To enable the reader design novel solutions for more efficiently managing the communication aspects that fall in the realm of current and future ADAS.

5.2 Introduction

As mentioned also in the previous chapter, the automotive world has been lately experiencing a trend related to the extensive use of ICT inside vehicles and in transportation infrastructures. The results of this trend are reflected on the term "ITS" (as mentioned above), which envisages systems that are either related to road infrastructures, making the infrastructure "intelligent," or used inside vehicles traveling on road, attributing vehicles with intelligence, forming the so called "ADAS", which envisage, e.g., a vehicle being equipped with an ITS that might be aided to avoid an emergency situation caused by another vehicle that has suddenly gone out of order, through V2V and V2I communication technologies. In this case, after gathering the necessary information, the vehicle's intelligent management system that is part of its ADAS informs the driver that he should slow down and potentially make a turn so as to avoid hazardous implications. Intelligence lies in the ITS's proactive decision-making functionality upon alternatives, which would be otherwise feasible only after the driver could see/identify the emergency.

G. Dimitrakopoulos, *Current Technologies in Vehicular Communications*,
DOI 10.1007/978-3-319-47244-7_5

Despite the recent advances in ADAS, there is still way to go for utilizing the appropriate management functionality for maximizing transportation efficiency and safety. The approach proposed herein aims to contribute towards this direction. This is justified as follows:

- Currently, the collection of context information, the solution of optimization problems and the application of reconfiguration decisions is an off-line process, applied in medium (or long) time scales. However, the traffic conditions that should be handled by vehicles may frequently change in a sudden or recurring manner. So, on one hand, traffic needs to be assessed in real time. On the other hand, traffic patterns resulting from a learning process could add accuracy to the messages communicated to the drivers; *in this context, novel management functionalities try to assess and exploit real-time traffic information through the (networks of) sensors and the associated decision-making algorithms.*
- Legacy traffic assessment and management systems are mainly centralized. Moreover, the communication among the central management entities and the vehicles is being done through internet, satellite, or cellular systems. Specifically, vehicles dispose positioning systems and obtain information on the traffic situation. The driver is thus capable of deciding on the proper direction to follow. This means that, in principle, such systems are complex, as well as unsuitable for adapting, in short time scales, to context changes. Novel approaches, in turn, *tend to operate in a completely autonomous manner, exchanging information amongst neighboring vehicles without any central control and policy-making entity.*
- Intelligence embedded in vehicles is still at a very low level and there is no assessment in the vehicle of the overall safety status that would rely on a correlation of the global traffic condition and the vehicle and driver behavior. As such, latest management functionalities *need to contribute to a significant increase in the vehicle's intelligence, through its valuable help and support that provides to the driver a priori.*

5.3 The Current Wireless Landscape: Towards Cognitive Systems

The wireless world has been lately migrating beyond the 4G era towards the 5G era. From a technological perspective, legacy (conventional) access network technologies, called "Radio Access Technology (RAT) standards," coexist and cooperate with currently emerging as well as completely new standards. In this respect, today's wireless world comprises numerous RATs of diverse nature, which can however be classified as follows.

- Wireless wide area networking technologies, which include, among others, mobile communications, as well as broadcasting technologies, such as DVB/DAB (Digital Video Broadcasting/Digital Audio Broadcasting).
- Wireless networking technologies of a shorter range that include, among others, wireless local and personal area networks (WLANs/WPANs), as well as wireless ad hoc networks and wireless sensor networks (WSNs).

Obviously, it is not possible to stand still. In the near future, even higher bit rates will be supported. But more significant than the bit rates are the capabilities of future networks that include full integration of IP, even smaller cells, self-planning dynamic topologies, flexible use of the spectrum, and utilization of precise user location. As to trends in services, it will become more and more important to deliver the right information at the right time and to the right place.

However, as can be seen day by day, novel communication systems become more and more complex. Complexity is usually derived (1) from the heterogeneous network and terminal infrastructure that needs to be tackled every time and (2) to the continuously increasing level of complexity of services and applications that arise from the ever-increasing user expectations for dependable, reliable, and secure services. The deployment of high complexity systems can be facilitated through several concepts, one of which is the reconfigurability concept, often seen as an evolution of "software defined radio."[113] Reconfigurability provides the technologies that are essential for terminals and network elements to dynamically (online) select and operate with those RATs that are considered as most appropriate for tackling the specific conditions encountered in a certain region and time zone. RATs are SDR based, i.e., they can be installed and uninstalled only through the appropriate (de)activation of the respective software components.

Moving one step further, complexity can be fought through the design of communication infrastructures on the premises of cognitive networking principles.[114] In general, "a cognitive system is capable of retaining knowledge from past interactions with the external environment and decide upon its future behavior based (1) on this knowledge, (2) other goals, and also (3) policies, so as to adapt to optimize its performance." It is anticipated that cognitive systems can facilitate the design, development, and integration of novel services and applications. An area of applications where cognitive systems could find prosper ground is transportation. Indicatively, a cognitive system placed inside a vehicle might seem like the one shown in Fig. 5.35.

As shown on the figure, the operation of a cognitive system placed inside a vehicle can be reflected on a feedback look. The system at time "t1" retrieves context information, potentially on traffic, velocity of neighboring vehicles, etc. Through the analysis of this information (at time "t2"), while taking into consideration its own preferences, goals, and policies, the system (at time "t3") decides on its actions, e.g., issue a directive towards the driver to change the vehicle's direction. The output of the system is stored on a "knowledge database," which might simply be a matrix, for future reference. This means that the system keeps track of its actions, so as to learn from their implications, in order to facilitate future decisions. This is repeated in a machine learning process[115] that leads to cognition.

[113] W. Hasselbring, R. Reussner, "Towards trustworthy software systems", IEEE Computer, Vol. 29, No. 4 April 2006

[114] Thomas R, Friend D, DaSilva L and McKenzie A. Cognitive networks: adaptation and learning to achieve end-to-end performance objectives. IEEE Commun. Mag., Vol. 44, No. 12, December 2006

[115] T. Mitchel, "Machine learning", McGraw-Hill, 1997

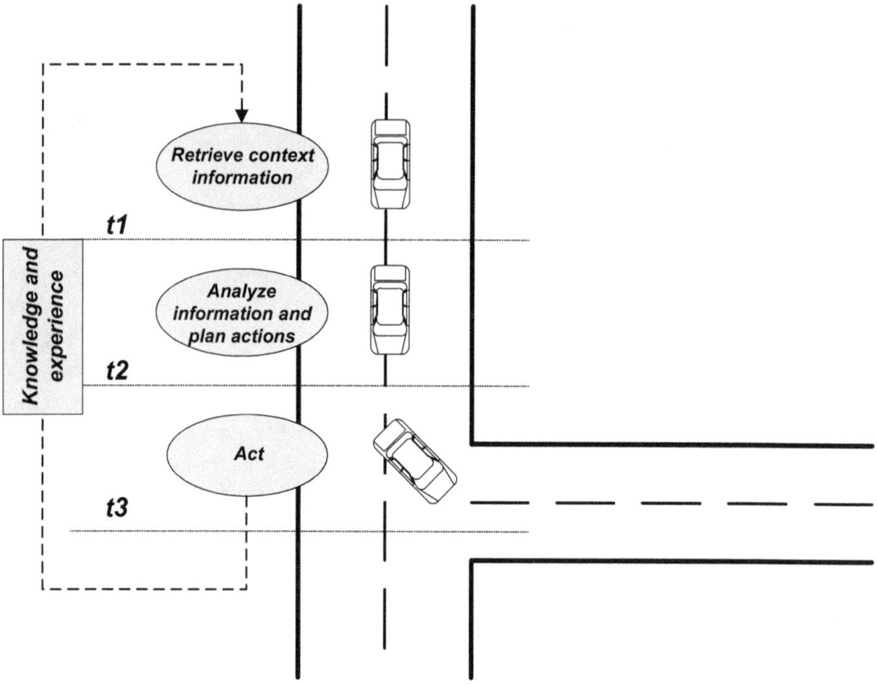

Fig. 5.35 Operation of a cognitive system

In general, it is believed that reconfigurable and cognitive systems move towards the most promising directions and technologies in the sense of removing any potential limitations that derive from business perspectives and providing the means to implement the vision of true end-to-end connectivity.

5.4 Wireless Sensor Networks (WSNs)

WSNs may comprise hundreds of nodes, which operate on the grounds of small batteries and thus its viability may depend on the resources consumption on behalf of its nodes. WSNs may exist, e.g., in areas where certain measurements need to take place, such as temperature, pressure, humidity, and velocity. Each time a sensor (node) receives a trigger, it forwards the relevant information to the whole network (WSN). Other sensors (nodes) receive this information and keep on forwarding it, up to a point where one or more "SINK nodes," i.e., nodes that have less energy limitations, larger processing power and thus undertake to gather any significant information (also further process it to an external network). The above are shown in Fig. 5.36.

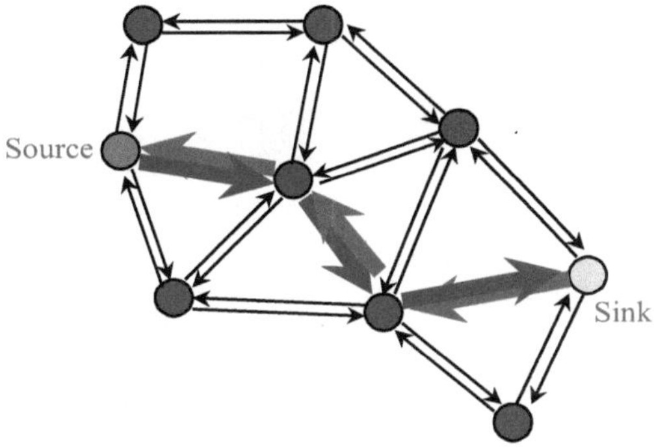

Fig. 5.36 Information transfer in a WSN

The way information is forwarded inside the WSN may vary, since there are many options for packet routing. What is more, nodes may be (intentionally) moving, while some nodes may cease to operate due to energy problems. The above render information transfer a tough problem.

In general, WSNs find many exciting applications. In this respect, the next subsection shows how WSNs can form part of intelligent transportation systems in conjunction with cognitive systems.

5.5 Cognitive Management Systems

This section discusses on an intelligent management functionality system for 4G/5G network segments operating in accordance with the cognitive networking paradigm. This corresponds to the less distributed part of the overall management architecture, as previously described. The approach presented may be short-term oriented (when semi-distributed, as explained in the sequel), since it sounds rather attractive for MNOs, due to its potentiality to permit a high level of network control. It still constitutes an approach that needs to be able to guarantee for an acceptable level of scalability, an aspect which is also of great importance.

5.5.1 General Characteristics

The general definition of cognitive networks implies some very advanced capabilities, which spring from the necessity to encompass reconfiguration (change in the behavior of the segment, reflected on parameters/infrastructure variations) features, enhanced by cognition capabilities. As part of the reconfiguration, at the PHY/MAC layers, there can be elements (hardware components, such as reconfigurable transceivers) that dynamically change the RATs they operate and the spectrum they use,

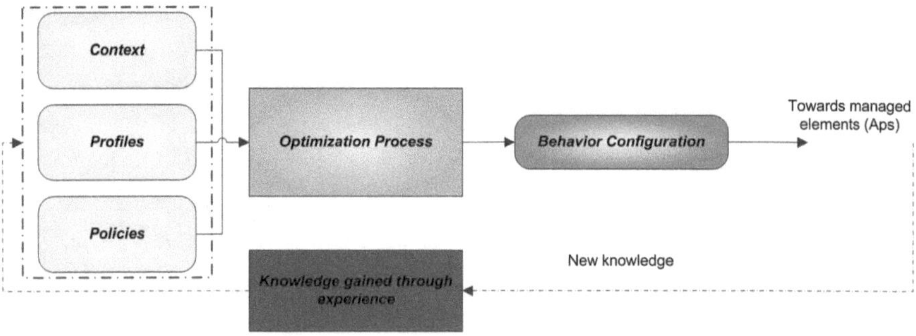

Fig. 5.37 Cognitive management functionality

in order to improve their QoS levels offered. On the other hand, such changes should be performed in the best possible way, i.e., the changes should be based on the applicable policies related to the context (combination of environment characteristics and specific event requirements) and on the selection of the most appropriate reconfiguration pattern. In general, this semi-distributed approach covers the motivation, requirements, functionality, and engineering challenges for a distributed functionality, which yields a powerful and scalable means that leads to cognitive, wireless access, networks. Furthermore, it includes the development of novel mechanisms for automating the procedure of deciding for the optimum reconfiguration, thus facilitating and optimizing the planning and management mechanisms.

Figure 5.37 provides the overall description of the management functionality proposed for managing a cognitive network segment.

The proposed management mechanisms undertake decisions that affect the protocol stack in a cross-layer fashion. The next subsections provide information on the problem's inputs (context, profiles, and policies) and output (configuration of behavior), as well as on the necessary cognition features that the functionality covers.

5.5.2 *Contextual Acquisition*

This part of the input is probably the most crucial, referring to the gist of the problem, due to its interactions with the environment, which constitute the primary reasons that urge a system to adapt to stimuli. Context information is monitored and discovered (sensed) for each element of the network segment and for its environment. Context information reveals the status of the elements, in the network segment, and of their environment (therefore performance, fault, etc. triggers will be covered). Context includes information on locations, time, traffic demand, mobility levels, interference conditions with managed elements, etc.

5.5.3 *Profiles Derivation*

This part provides information on the candidate configurations of the elements of the segment, such as the set of transceivers of each element, the set of operating RATs, as well as the set of spectrum carriers. Moreover, this part also describes the profiles (e.g., preferences, requirements, constraints) of user classes, applications and terminals, etc.

5.5.4 *Policies Extraction*

Policies designate rules and functionality that should be followed in context handling. Specifically, this part provides information on the MNO policies with respect to reconfiguration strategies, i.e., MNO preferences and priorities on goals to be achieved. These are related to the maximization of the QoS levels (performance, availability, reliability), and the minimization of cost factors (resource consumption). Furthermore, this part provides information on MNO agreements with cooperative MNOs. Policies and goals should guarantee for the maximum possible level of stability.

5.5.5 *Output*

The output provides actions that will determine the behavior of the network segments (configurations at all layers of the protocol stack of each element of the segment).

At the physical and MAC (Medium Access Control) there will be the selection of RATs (Radio Access Technologies), spectrum, transmission power, as well as other parameters and algorithms depending on the network element and terminal capabilities.

At the network level there will be actions related to element interconnection (legacy or emerging network topologies like mesh), routing, and congestion control

At the application level, there will be actions related to the allocation of applications to QoS levels.

5.5.6 *Cognitive Features*

Cognitive management functionality constitutes an essential step towards manners of planning and managing purely cognitive network segments. This is depicted on its input, output, and the approached optimization techniques. Specifically, the proposed functionality is applicable at various levels of distribution, ranging from semi-distributed to fully distributed schemes. The semi-distributed schemes are a first step for introducing cognition in 4G/5G wireless communications, while they can also serve as benchmarks for the fully distributed approaches, which can support and ease cognition, due to their lower level of dependence on lateral factors. Solution algorithms for this part of the overall management functionality for cognitive networks can be based on optimization techniques, in addition to machine learning functionality and artificial intelligence techniques. In fact, machine learning functionality and artificial intelligence algorithms attribute the management functionality with knowledge and experience (at least) in the areas described below and can qualify the functionality in the cognitive domain.

Context information is obtained through interactions with the environment, which lead to reasoning and perception, through appropriate machine learning techniques. The network segment is thus able to gain knowledge from those interactions and be aware of the optimum behavior for various contexts.

The constant updates in the information provided by the "profiles" part of the functionality leads to significant knowledge gains with respect to user behavior. This is essential in the quest for seamless provisioning of services and unparalleled quality, tailored to individual user needs. This implies that the process of serving users is facilitated and optimized through experience.

Cognitive features lie also in the "policies" part of the management functionality. Specifically, the suitability and efficiency of different policies possesses great importance in handling versatile contextual situations. Consequently, learning the most optimum policy and the most appropriate goals to be achieved may become valuable for NOs in successfully (transparently, fast, and securely) handling difficult conditions.

5.6 Management Functionality Approaches for ADAS

5.6.1 High-Level Approach

ADAS-oriented management functionality may benefit significantly if designed on the premises of cognitive networking principles, in order to exploit the (collective) intelligence accumulated through the exchange of information amongst numerous vehicles that lie in a certain vicinity.

In general, this can be achieved through networks of sensors dynamically formed by vehicles that fall at a certain geographical range and capable of allowing communication among the different network nodes (vehicles). This in turn leads to important information exchange amongst the network vehicles. Such information may not only improve the quality of transportation in terms of reduction of traffic congestions, but may also be important in automatically reducing accident risks and emergency situations.

Such functionality is shown at a high level in Fig. 5.38 and presented in detail in the next section.

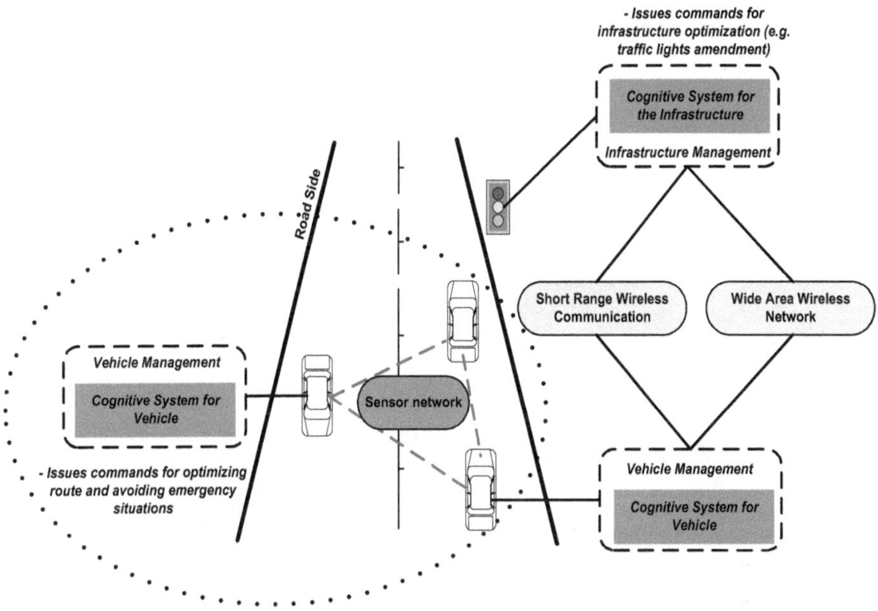

Fig. 5.38 High-level view of functionality

5.6.2 Requirements

The most significant goal of intelligent management functionality for vehicles, as well as for the whole transportation infrastructure, is to improve the levels of efficiency and safety of mobility. In this respect, there are several lateral requirements that need to be tackled, before the design and development process. Those are outlined below:

- Awareness of contextual situations, so as to identify the current context and help the vehicle adapt to it dynamically, securely and fast and, as such, provide the maximum possible levels of quality.
- Personalization, so as not only to support various classes of vehicles/drivers, but also to provide solutions tailored to the individual driver profile.
- Support of pervasive computing, so as to enable the existence and operation of sensors, ad hoc networking entities, and also local area networks in all application areas.
- Always-best connectivity for providing seamless network access for enabling the functionality's operation in heterogeneous environments.
- Collaboration with alternate RATs for enabling the seamless operation of the functionality.
- Scalability, i.e., the ability to provide solutions at various levels, so as to be able to act either in a collaborative or in an autonomic manner, depending on the specific needs.

A solution that derives from the analysis of the aforementioned requirements, for the cognitive functionality for the management of vehicles, is presented in the following.

5.6.3 Indicative Architecture and Description of Components

The architecture of the proposed functionality is shown in Fig. 5.39. As can be shown in the figure, the functionality comprises the following complementary components.

5.6.4 Vehicle Sensors and WSNs

Vehicle sensors are capable of enabling the ad hoc formation of networks (WSNs) amongst neighboring vehicles, in order to allow the communication and information exchange. In general, a vehicle may comprise several sensors. The information exchanged among them can be classified as follows:

(a) High-level data. This type of information includes knowledge on the congestion level, alert regarding potential emergencies, characterization of driver's behavior, characterization of vehicle's overall condition (good, normal, bad) and cruising behavior, information on the road condition (e.g., slippery), information on neighboring vehicles and their cruising behavior, general knowledge on location (mountain road, city road, etc.).

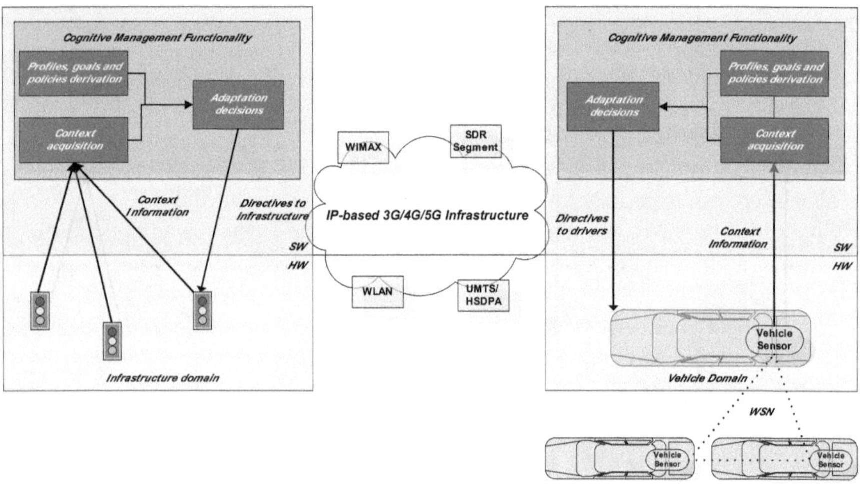

Fig. 5.39 Architecture of proposed functionality for ADAS

(b) Low-level data. This type of information includes information on the vehicles, such as their accurate positions (distance among them), their velocities and directions, their capabilities braking distances, accelerations. Furthermore, information on the drivers is also included, such as the driver's profiles, driving habits, capabilities, and preferences (policies). The resulting information can be used in the inference of a driver's current tasks as well as driver modeling; that is, common driving-related tasks are recognized by the interpretation of sensor measurements of vehicle–driver interactions (e.g., operating of the foot pedals, gear changes, pressing buttons on the instrument panel, or looking into mirrors) as well as state information about the vehicle (e.g., location, fuel levels, engine status). This process is performed unobtrusively, without disturbing the driver in his natural behavior. The focus is on recognizing short-term tasks that a driver may pursue.

In general, the sensors are required to decide on how to process in-vehicle data, which aggregated data are to be sent, how often, etc. Sensor measurements are processed in a hierarchical manner with specialized reasoning techniques, which yield information about the vehicle–driver interactions at various abstraction levels.

5.6.5 Vehicle Cognitive Management Functionality (V-CMF)

Input. The V-CMF input includes contextual information acquired from the sensors and the WSNs, regarding the status of the vehicle, its velocity, direction, neighboring vehicles' positions, directions and velocities, road side information such as road condition, congestion levels and potential emergencies, as well as traffic lights and road signs conditions. Moreover, the input includes information on the driver's profiles. To do so, a predefined set of driver states is inferred from interpreted driver monitoring data (this information is also retrieved from the sensors). Moreover, plan recognition techniques are explored to derive driver state and behavior. This means that sequences of interactions between the driver and the vehicle, the raw signals about driver's physical condition (eye blink frequency, eyelid opening, head movement, profile, operating the foot pedals, pressing buttons on the instrument panel, steering wheel activity, etc.) as well as vehicle state information are also acquired in the form of a facial driver recognition, which allows for the detection of differences between changing driving styles. Finally, driver's goals, priorities, and policies are also included. Goals and policies aim at maximizing the performance, safety, reliability, and stability of the decisions taken, from an end-to-end perspective.

Output. The V-CMF results in issuing commands (directives) towards the driver, so as to adapt the vehicle's road behavior, and tackle any emergency situations, through emergency braking, or through vehicle direction correction (again based on perception and reasoning). Moreover, congestion can be avoided through the reconsideration of the vehicle's advisable route.

Decision-making. Several approaches can be envisaged for the decision-making process. In general, the V-CMF utilizes appropriate intelligent algorithms that exploit the input in terms of optimizing an objective function (OF)[116] that refers to certain aspects of the vehicle's behavior (overall delay, mean velocity, etc.).

Knowledge and experience. The information acquired is processed and appropriately interpreted, so as to infer knowledge and experience. The knowledge model aims at capturing various aspects, such as the driver's state (attentiveness, fatigue) and behavior, as well as the overall vehicle environment. Moreover, information on certain contextual situations (recurrent or emergencies) and the way they have been confronted is retained, so as to serve for future decisions.

5.6.6 *Infrastructure Cognitive Management Functionality (I-CMF)*

Input. The I-CMF acquires input from the WSNs regarding the condition of elements or segments of the transportation infrastructure (traffic lights, road signs, road conditions, congestion levels, overall load in telecommunications network), so as to be aware of the current context. Moreover, the input includes information on the vehicle's profiles, as well as goals and policies dictated by the transportation authorities.

Output. The I-CMF is targeted at deciding on the proper configuration of small elements or larger segments of the transportation infrastructure, i.e., traffic lights and road signs.

Decision-making. Several optimization algorithms are envisaged for the I-CMF. Those algorithms are targeted at achieving optimal performance, safety, reliability, and stability, from the end-to-end perspective. Additionally, cost factors are also addressed, through the minimization of the overall load in the telecommunication network.

Knowledge and experience. Decisions need to be enhanced with learning capabilities, so as to accelerate and improve the efficiency of the necessary adaptation (reconfiguration) actions. In particular, algorithms are enriched with knowledge features, through the incorporation of basic learning techniques, such as pattern matching and context recognition that help a system compare current contextual situations with past ones and identify already applied solutions that could be put into effect. However, faster, more effective, and more stable leaning strategies can also be adopted.

It should be noted that distributed context acquisition and decision-making at various degrees of distribution are enabled by the proposed architecture, i.e., either autonomously from the V-CMF or in a collaborative manner from the I-CMF. In general, the aforementioned components cooperate with each other, so as to generate knowledge

[116] G. Dimitrakopoulos, P. Demestichas, K. Tsagkaris, A. Saatsakis K. Moessner, M. Muck, D. Bourse "Emerging Management Concepts for Introducing Cognition in the Wireless, B3G World", Wireless Personal Communications, Vol. 48, Issue 1, pp. 33–47, January 2009

from various sources and result in useful directives towards the vehicles and the transportation infrastructure. The way this is realized is described in the next section.

5.6.7 Indicative Information Flow

This section aims at exemplifying the usual operation approaches of such functionality. In this respect, Fig. 5.40 depicts a scenario for showcasing the exchange of information among the functionality's components.

The scenario is twofold, in the sense that it comprises (1) a part that is tackled by the V-CMF and (2) a part that is tackled by the I-CMF. In both parts, the initial trigger is supposed to originate in the WSNs. The WSNs exchange information among their nodes (sensors).

In the first part of the scenario, information that is derived from the WSNs, i.e., context information along with vehicle's and driver's profiles, goals, and policies, is transferred to the V-CMF. The V-CMF gathers this information and compares it with data already retained in its "knowledge database," so as to compare the current context (and the relevant decisions taken) with past ones, and thus utilize past knowledge and experience, before taking any decisions. Then, the V-CMF runs an optimization process, as described above, so as to take decisions and issue any necessary commands (directives) to the driver, valuable in context handling.

In the second part of the scenario, again WSNs send information on the current contextual situation to the I-CMF. A main difference here is that the I-CMF gathers also information from the transportation infrastructure, regarding the condition of the elements (traffic lights, road signs, etc.) and larger segments (road sides) of the infrastructure. What is more, the I-CMF gathers information on profiles, goals, and policies from the transportation authorities, which constitute rules that need to be

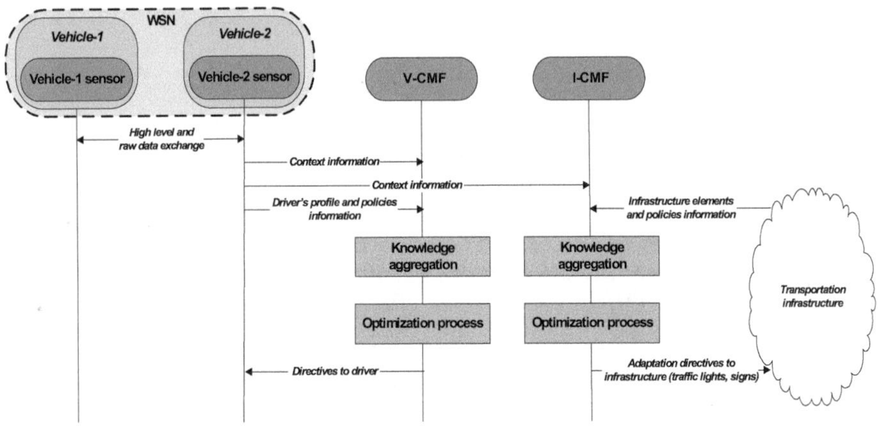

Fig. 5.40 Functionality components and indicative information flow

taken into consideration by the I-CMF. The I-CMF then compares the information gathered with past contexts confronted and the associate decisions taken. Finally, the I-CMF runs its own optimization process and results in decisions regarding the configurations of the infrastructure, so as to optimize certain criteria, such as resources consumption, resolution of an incident, and minimization of cost.

5.7 Conclusions

ADAS are accompanied by management functionality that facilitates decision-making that makes the driver's and passengers' life easier and safer.

In this respect, this chapter has presented some approached that are researched in the international literature, concerning decision-making systems as part of ADAS. Such approaches lately tend to utilize knowledge and experience, thus being turned into cognitive ones. We will see later on an important link of cognitive approaches to autonomous driving, since autonomous vehicles benefit from knowledge and experience through a machine learning process, so as to undertake the role of the driver whenever needed.

5.8 Review Questions

Question 5.1:
Can you give an overview of the history of generations within the wireless communications landscape?

Question 5.2:
Why the increase in ICT advances incurs higher complexity with regard to management systems and how can complexity be fought?

Question 5.3:
What is the difference between adaptability and reconfigurability?

Question 5.4:
What do cognitive (knowledge-based) networks bring as an asset to any network/system?

Question 5.5:
Can you think of an analogy between the cognitive loop and other areas of human initiative (e.g., health)?

Question 5.6:
How can cognitive management systems facilitate the operation and utilization of ADAS?

Question 5.7:
What is the difference between cognitive management functionality targeted for vehicles and the respective functionality targeted for the infrastructure?

Question 5.8:
What is the benefit of cognitive networks in the case knowledge and experience cannot be appropriately exploited in decision-making?

Question 5.9:
Can you provide three examples of cognitive management functionality in the transport sector?

Question 5.10:
Is knowledge-based interaction appropriate for large-scale decision-making in urban transport?

Chapter 6
The Future: Towards Autonomous Driving

6.1 Goal of Chapter

- To enable the reader understand the differences between highly automated and autonomous driving.
- To familiarize the reader with the fundamentals of autonomous driving.
- To set the scene for future research on vehicular communications, enabled by autonomous driving applications.
- To present the advantages and drawbacks of autonomous driving, so as to enable the reader gain a holistic knowledge upon it and be able to contribute to its future developments.

6.2 Highly Automated Driving

Automated driving is the conceptual extension of the driver assistance system and cooperative mobility, when automation takes the whole control of the vehicle. This concept has been studied for years by many researchers.

One must speak of automated highways in plural. Indeed, it can mean automated shuttles for door-to-door short distance transportation, automated corridors (mainly highways and suburban roads) dedicated to trucks or to light vehicles, partially or fully automated vehicles to deal with traffic jam or for networks that cannot be instrumented such that the secondary road networks.

Automated shuttles were the first achievements. INRIA with its Cycab vehicle is a pioneer in the field (1993). This concept has been demonstrated most recently in real operating conditions through Europeans projects such as Cybercars (2004),

G. Dimitrakopoulos, *Current Technologies in Vehicular Communications*,
DOI 10.1007/978-3-319-47244-7_6

Cybercars-2 (2008), and CityMobil (end 2012).[117] Other teams worldwide are now working on such concepts. This is the case in France for the LASMEA through National PREDIT projects MobiVIP (2007), CityVIP (2011) and with his new VIPA vehicle (2010). There are also achievements in Switzerland with the Serpentine system (2001) or in Nederland with the FROG Parkshuttle. The first commercial use was conducted by this company to deserve the P3car parking at the Amsterdam Schiphol Airport (from December 1997).

Researches also address trucks platooning. That was the case within the Chauffeur1 and Chauffeur2 (2003) projects coordinated by Daimler in order to design corridors dedicated to these vehicles. In 1997 Path program demonstrations have demonstrated the automated vehicles platooning concept in order to increase road capacity. The concept of a suburban roads scenario has been particularly studied in the book « la route automatisée: un scenario périurbain ». The DARPA Grand Challenges (2004 and 2005) and Urban Challenge (2007) have demonstrated the feasibility of fully automated vehicles. The CMU and Berkeley American universities made significant achievements in these challenges. Currently the University of Parma is undertaking another challenge (VIAC: the Vislab Intercontinental Autonomous Challenge) by connecting Parma to Shanghai (13,000 km) in 3 months with a first driven vehicle followed by two electric unmanned vehicles. Active research projects are SARTRE and HAVE-IT.[118]

6.3 Autonomous Driving

6.3.1 Introduction

In general, "autonomous" means having the power for self-governance.[119] There have been several research and development projects dedicated to vehicle autonomy. However, most of them have in fact only been automated (made to be automatic) due to a heavy reliance on artificial hints in their environment, such as magnetic strips. Autonomous control implies good performance under significant uncertainties in the environment for extended periods of time and the ability to compensate for system failures without external intervention. As can be seen from many projects mentioned, it is often suggested to extend the capabilities of an autonomous car by implementing communication networks both in the immediate

[117] Yoann Pigné, Grégoire Danoy and Pascal Bouvry: A Platform for Realistic Online Vehicular Network Management: In proceedings of the IEEE MENS Workshop at Globecom 2010: 6–10 December 2010, Miami, Florida, USA

[118] C. Shen, D. Pesch, J. Irvine: "A Framework for Self-Management of Hybrid Wireless Networks Using Autonomic Computing Principles", 3rd Annual Communication Networks and Services Research Conference CNSR 2005, pp. 261–266.

[119] Antsaklis, Panos J.; Passino, Kevin M.; Wang, S.J. (1991)."An Introduction to Autonomous Control Systems" (PDF). IEEE Control Systems 11 (4): 5–13. doi:10.1109/37.88585.

vicinity (for collision avoidance) and far away (for congestion management). By bringing in these outside influences in the decision process, some would no longer regard the car's behavior or capabilities as autonomous.

For example, Wood[120] writes, "This Article generally uses the term 'autonomous,' instead of the term 'automated.' The term 'autonomous' was chosen because it is the term that is currently in more widespread use (and thus is more familiar to the general public). However, the latter term is arguably more accurate. 'Automated' connotes control or operation by a machine, while 'autonomous' connotes acting alone or independently. Most of the vehicle concepts (that we are currently aware of) have a person in the driver's seat, utilize a communication connection to the Cloud or other vehicles, and do not independently select either destinations or routes for reaching them. Thus, the term 'automated' would more accurately describe these vehicle concepts".

In the United States, the National Highway Traffic Safety Administration (NHTSA) has proposed a formal classification system[121]:

- Level 0: The driver completely controls the vehicle at all times.
- Level 1: Individual vehicle controls are automated, such as electronic stability control or automatic braking.
- Level 2: At least two controls can be automated in unison, such as adaptive cruise control in combination with lane keeping. Example: Tesla Model S.
- Level 3: The driver can fully cede control of all safety-critical functions in certain conditions. The car senses when conditions require the driver to retake control and provides a "sufficiently comfortable transition time" for the driver to do so.
- Level 4: The vehicle performs all safety-critical functions for the entire trip, with the driver not expected to control the vehicle at any time. As this vehicle would control all functions from start to stop, including all parking functions, it could include unoccupied cars.

An alternative classification system based on five different levels (ranging from driver assistance to fully automated systems) has been published by SAE, an automotive standardization body.[122]

6.3.2 Advantages

Autonomous vehicles hold significant advantages. It is expected that an increase in the use of autonomous vehicles would make possible such benefits as:

---

[120] Wood, S. P.; Chang, J.; Healy, T.; Wood, J. "The potential regulatory challenges of increasingly autonomous motor vehicles.". 52nd Santa Clara Law Review 4 (9): 1423–1502.

[121] "U.S. Department of Transportation Releases Policy on Automated Vehicle Development". National Highway Traffic Safety Administration. 30 May 2013. Retrieved 18 December2013.

[122] "AdaptIVe system classification and glossary on Automated driving" (PDF).

1. Avoid traffic collisions caused by human driver errors such as reaction time, tail gating, rubbernecking, and other forms of distracted or aggressive driving.
2. Increased roadway capacity and reduced traffic congestion due to reduced need for safety gaps and the ability to better manage traffic flow.
3. Relief of vehicle occupants from driving and navigation chores.
4. Higher speed limit for autonomous cars.
5. Removal of constraints on occupants' state—in an autonomous car, it would not matter if the occupants were under age, over age, unlicensed, blind, distracted, intoxicated, or otherwise impaired.
6. Reduction of physical space required for vehicle parking, and vehicles will be able to drive where space is not scarce.
7. Reduction in the need for traffic police and premium on vehicle insurance.
8. Reduction of physical road signage—autonomous cars could receive necessary communication electronically (although physical signs may still be required for any human drivers).
9. Smoother ride.
10. Reduction in car theft, due to the vehicle's increased awareness.
11. Increased ergonomic flexibility in the cabin, due to the removal of the steering wheel and remaining driver interface, as well as no occupant needing to sit in a forward-facing position.
12. Increased ease-of-use of large vehicles such as motorhomes.
13. When used for car-sharing
14. Reduces total number of cars.
15. Enables new business models such as mobility as a service which aim to be cheaper than car ownership by removing the cost of the driver.
16. Elimination of redundant passengers—the robotic car could drive unoccupied to wherever it is required, such as to pick up passengers or to go in for maintenance.

6.3.3 Disadvantages and Obstacles

In spite of the various benefits to increased vehicle automation, some foreseeable challenges persist:

1. Liability placed on manufacturer of device and/or software driving the vehicle.
2. Time needed to turn an existing fleet of vehicles from nonautonomous to autonomous.
3. Resistance by individuals to forfeit control of their cars.
4. Implementation of legal framework and establishment of government regulations for self-driving cars.
5. Inexperienced drivers if complex situations require manual driving.
6. Loss of driving-related jobs.
7. Resistance from professional drivers and unions who perceive job losses.

8. Loss of privacy. Sharing of information through V2V (Vehicle to Vehicle) and V2I (Vehicle to Infrastructure) protocols.
9. Self-driving cars could potentially be loaded with explosives and used as bombs.
10. Ethical problems in situations where an autonomous car's software is forced during an unavoidable crash to choose between multiple harmful courses of action.
11. Current police and other pedestrian gestures and nonverbal cues are not adapted to autonomous driving.
12. Software reliability.
13. A car's computer could potentially be compromised, as could a communication system between cars by disrupting camera sensors, GPS jammers/spoofing.
14. Susceptibility of the car's navigation system to different types of weather.
15. Autonomous cars may require very high quality specialized maps to operate properly. Where these maps may be out of date, they would need to be able to fall back to reasonable behaviors.
16. Competition for the radio spectrum desired for the car's communication.
17. Current road infrastructure may need changes for autonomous cars to function optimally.

6.3.4 *Legislation and Political Decisions*

In Europe, cities in Belgium, France, Italy, and the UK are planning to operate transport systems for driverless cars,[123],[124],[125] and Germany, the Netherlands, and Spain have allowed testing robotic cars in traffic. In 2015, the UK Government launched public trials of the LUTZ Pathfinder driverless pod in Milton Keynes.[126] Since Summer 2015 the French government allowed PSA Peugeot-Citroen to make trials in real conditions in the Paris area. The experiments will be extended to other French cities like Bordeaux and Strasbourg by 2016.[127] The alliance between the French companies THALES and Valeo (provider of the first self-parking car system that equips Audi and Mercedes premi) is also testing its own driverless car system.[128]

---

[123] "Driverless cars take to the road". E.U.CORDIS Research Program CitynetMobil. Retrieved April 15, 2016.

[124] "Snyder OKs self-driving vehicles on Michigan's roads". Detroit News. 27 December 2013. Retrieved April 15, 2016.

[125] "BBC News—UK to allow driverless cars on public roads in January". BBC News. Retrieved April 15, 2016.

[126] Burn-Callander, Rebecca (11 February 2015). "This is the Lutz pod, the UK's first driverless car". Daily Telegraph. Retrieved April 15, 2016.

[127] "Autonomous vehicle: the automated driving car of the future". PSA PEUGEOT CITROËN.

[128] Valeo Autonomous iAV Car Driving System CES 2015.YouTube. 5 January 2015.

In the United States, state vehicle codes generally do not envisage—but do not necessarily prohibit—highly automated vehicles.[129] To clarify the legal status of and otherwise regulate such vehicles, several states have enacted or are considering specific laws. As of the end of 2013, four US states, (Nevada, Florida, California, and Michigan), along with the District of Columbia, have successfully enacted laws addressing autonomous vehicles.

In June 2011, the Nevada Legislature passed a law to authorize the use of autonomous cars. Nevada thus became the first jurisdiction in the world where autonomous vehicles might be legally operated on public roads. According to the law, the Nevada Department of Motor Vehicles (NDMV) is responsible for setting safety and performance standards and the agency is responsible for designating areas where autonomous cars may be tested.[130,131,132] This legislation was supported by Google in an effort to legally conduct further testing of its Google driverless car.[133]

The Nevada law defines an autonomous vehicle to be "a motor vehicle that uses artificial intelligence, sensors and global positioning system coordinates to drive itself without the active intervention of a human operator." The law also acknowledges that the operator will not need to pay attention while the car is operating itself. Google had further lobbied for an exemption from a ban on distracted driving to permit occupants to send text messages while sitting behind the wheel, but this did not become law.[134,135] Furthermore, Nevada's regulations require a person behind the wheel and one in the passenger's seat during tests

6.3.5 *The Way to the Future*

Autonomous vehicles are still a developing technology; a large number of companies and researchers have speculated about future developments and the possible effects of the cars. In 2014 Raj Rajkumar, director of autonomous driving research at Carnegie-Mellon University, said that the artificial intelligence necessary for a

---

[129] Bryant Walker Smith (1 November 2012). "Automated Vehicles Are Probably Legal in The United States". The Center for Internet and Society (CIS) at Stanford Law School. Retrieved April 15, 2016.

[130] "Nevada enacts law authorizing autonomous (driverless) vehicles". Green Car Congress. 25 June 2011. Retrieved April 15, 2016.

[131] Alex Knapp (22 June 2011). "Nevada Passes Law Authorizing Driverless Cars". Forbes. Archived from the original on 28 June 2011. Retrieved 25 June 2015.

[132] Christine Dobby (24 June 2011). "Nevada state law paves the way for driverless cars". Financial Post. Retrieved 25 June 2015

[133] John Markoff (10 May 2011). "Google Lobbies Nevada To Allow Self-Driving Cars". The New York Times. Retrieved April 10, 2016

[134] "Nevada Passes Law Allowing Self-Driving Cars". Motor Trend. Retrieved 15 April 2016.

[135] "Nevada issues Google first license for self-driving car". Las Vegas Sun. Retrieved 15 April 2016.

driverless car would not be available “anytime soon” and that Detroit car makers believe “the prospect of a fully self-driving car arriving anytime soon is ‘pure science fiction’.”

Although the future of autonomous driving seems far, possible developments include the following:

- By 2017, Mobileye expects to release autonomous capabilities for country roads.
- By 2018, Mobileye expects to release autonomous capabilities for city traffic as well as capabilities where driver is not required to be alert.
- By 2018, Nissan anticipates to have a feature that can allow the vehicle maneuver its way on multilane highways.
- By 2018, Baidu plans to release its fully autonomous system.
- By 2018, Elon Musk expects Tesla Motors to have developed mature serial production version of fully self-driving cars, where the driver can fall asleep. However, he expects they would be allowed only some years after that, due to regulatory issues.
- By 2020, Volvo envisages having cars in which passengers would be immune from injuries. Volvo also claims vehicles will effectively be “crash free.”
- By 2020, Audi, BMW, Daimler, Ford, GM, Google, Kia, Mercedes-Benz, Nissan, Renault, Tesla, and Toyota all expect to sell vehicles that can drive themselves at least part of the time.
- By 2020, Google autonomous car project head’s goal to have all outstanding problems with the autonomous car be resolved.
- By 2024, Jaguar expects to release an autonomous car.
- By 2025, most new GM vehicles will have automated driving functions as well as vehicle-to-vehicle communication technology.
- By 2035, IHS Automotive report says will be the year most self-driving vehicles will be operated completely independently from a human occupant’s control.
- By 2035, Navigant Research forecasts that autonomous vehicles will gradually gain traction in the market over the coming two decades and by 2035, sales of autonomous vehicles will reach 95.4 million annually, representing 75 % of all light-duty vehicle sales.
- By 2040, expert members of the Institute of Electrical and Electronics Engineers (IEEE) have estimated that up to 75 % of all vehicles will be autonomous.

6.4 Conclusions

An autonomous—or self-driving—car is one that can accelerate, brake, and steer itself. Such cars have long been part of a utopian vision of the future because they will free people from the boring aspects of driving and open up exciting new ways to travel. The many attempts at realizing this vision over the years have been limited by the technology available. Now we are able to make autonomous cars a reality.

Autonomous driving has the power to change the world as we know it forever. This change will take place step by step, however, to ensure that the technology fits around how and where people use it.

Today, several vehicle manufacturers use this technology to create semiautonomous cars that make your journey easier and safer, while leaving you fully in control. For example, Mercedes-Benz, Volvo, and BMW have already the all-technologies keeping you a set distance from the car in front and in lane, at speeds up to 100–130 km/h.

The fully autonomous car goes further. It is able to perform all driving functions without supervision of the driver.

In between these two is the highly autonomous car. This technology will give you the option of handing over control—and responsibility—to the car on specific roads. You will be able to use your time as you choose, taking back control to enjoy driving whenever you like.

The similarities and differences among highly autonomous/highly automated and autonomous driving are shown on the table below.

	Semiautonomous driving	Autonomous driving
How it works	The car can drive itself (accelerate, brake, and steer) to a limited extent, i.e., supports the driver with keeping the distance to the vehicle in front and keeping the car in the lane in speeds up to 80 mph/130 km/h. However the driver remains responsible for monitoring, supervision, and over all operation of the vehicle and is expected to actively participate in the driving	The car drives itself (accelerates, brakes, and steers), and the driver is not responsible for monitoring, supervision, and over all operation of the vehicle
Responsibility	The driver is always legally responsible for driving the vehicle ("driver in the loop," "hands on the wheel, eyes on the road, mind on driving")	The driver will not be responsible for driving the vehicle when in autonomous mode ("driver out of the loop")
Benefits to you	The customer benefits a reconvenience, peace of mind and feeling of control	The customer benefit will be the freedom to engage in other activities (relax, create, entertainment, etc.) in autonomous mode
Legal status	This is currently legal, since it does not change the basic assumption of (licenced) driver always being responsible	There is currently no legal framework (except for testing, in certain jurisdictions)
Roll out	These features have launched and will continue to launch in increments (certain speeds, certain scenarios, etc.)	This may launch in increments (certain speeds, scenarios, conditions, applications, markets)

6.5 Review Questions

Question 6.1:
Is highly automated driving the same with having a very advanced ADAS?

Question 6.2:
What is the difference between highly automated driving and autonomous driving?

Question 6.3:
What is the level of autonomy already available in commercial vehicles today?

Question 6.4:
Do you think that autonomous driving will be the end of research in ADAS?

Question 6.5:
What are the ethics associated with autonomous driving, especially in cases of collisions with a not apparent liability?

FSC
www.fsc.org
MIX
Papier | Fördert gute Waldnutzung
FSC® C083411

Kolibri 360 GmbH
Ferdinand-Jühlke-Straße 7
99095 Erfurt, Deutschland
produktsicherheit@kolibri360.de